T0091750

PARTICLES IN THE EARLY UNIVERSE

EARLY UNIVERSE

High-Energy Limit of the Standard Model from the Contraction of Its Gauge Group

PARTICLES IN THE EARLY UNIVERSE

High-Energy Limit of the Standard Model from the Contraction of Its Gauge Group

Nikolai A Gromov

Russian Academy of Sciences, Russia

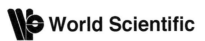

World Scientific

NEW JERSEY · LONDON · SINGAPORE · BEIJING · SHANGHAI · HONG KONG · TAIPEI · CHENNAI · TOKYO

Published by

World Scientific Publishing Co. Pte. Ltd.

5 Toh Tuck Link, Singapore 596224

USA office: 27 Warren Street, Suite 401-402, Hackensack, NJ 07601

UK office: 57 Shelton Street, Covent Garden, London WC2H 9HE

British Library Cataloguing-in-Publication Data
A catalogue record for this book is available from the British Library.

PARTICLES IN THE EARLY UNIVERSE
High-Energy Limit of the Standard Model from the Contraction of Its Gauge Group

Copyright © 2020 by World Scientific Publishing Co. Pte. Ltd.

All rights reserved. This book, or parts thereof, may not be reproduced in any form or by any means, electronic or mechanical, including photocopying, recording or any information storage and retrieval system now known or to be invented, without written permission from the publisher.

For photocopying of material in this volume, please pay a copying fee through the Copyright Clearance Center, Inc., 222 Rosewood Drive, Danvers, MA 01923, USA. In this case permission to photocopy is not required from the publisher.

ISBN 978-981-120-972-7

For any available supplementary material, please visit
https://www.worldscientific.com/worldscibooks/10.1142/11537#t=suppl

Typeset by Stallion Press
Email: enquiries@stallionpress.com

Preface

Group-Theoretical Methods are an essential part of modern theoretical and mathematical physics. The most advanced theory of fundamental interactions, namely Standard Model, is a gauge theory with gauge group $SU(3) \times SU(2) \times U(1)$. All types of classical groups of infinite series: orthogonal, unitary and symplectic as well as inhomogeneous groups, which are semidirect products of their subgroups, are used in different areas of physics. Euclidean, Lobachevsky, Galilean, Lorentz, Poincaré, (anti) de Sitter groups are the bases for space and space-time symmetries. Supergroups and supersymmetric models in the field theory predict the existence of new supersymmetric partners of known elementary particles. Quantum deformations of Lie groups and Lie algebras lead to non-commutative space-time models (or kinematics).

Contractions of Lie groups is the method for receiving new Lie groups from the initial ones. In the standard E. Wigner and E. Inönü approach, [Inönü and Wigner (1953)] continuous parameter ϵ is introduced in such a way that in the limit $\epsilon \to 0$ group operation is changed but Lie group structure and its dimension are conserved. In general, a contracted group is a semidirect product of its subgroups. In particular, a contraction of semisimple groups gives non-semisimple ones. Therefore, the contraction method is a tool for studying non-semisimple groups starting from the well known semisimple (or simple) Lie groups.

The method of contractions (limit transitions) was extended later to other types of groups and algebras. Graded contractions [de Montigny and Patera (1991); Moody and Patera (1991)] additionally conserve grading of Lie algebra. Lie bialgebra contractions [Ballesteros, Gromov, Herranz, del Olmo and Santander (1995)] conserve both Lie algebra structure and cocommutator. Contractions of Hopf algebras (or quantum groups) are introduced in such a way [Celeghini, Giachetti, Sorace and Tarlini (1990, 1992)] that in the limit $\epsilon \to 0$ new expressions for coproduct, counit and antipode appear which satisfies Hopf algebra axioms. All these give rise to the following generalization of the notion of group contraction on contraction of algebraic structures [Gromov (2004)].

Definition 0.1. Contraction of algebraic structure $(M, *)$ is the map ϕ_ϵ dependent on parameter ϵ

$$\phi_\epsilon : (M, *) \to (N, *'), \qquad (0.1)$$

where $(N, *')$ is an algebraic structure of the same type, which is isomorphic in $(M, *)$ when $\epsilon \neq 0$ and non-isomorphic when $\epsilon = 0$.

There is another approach [Gromov (2012)] to the description of non-semisimple Lie groups (algebras) based on their consideration over Pimenov algebra $P_n(\iota)$ with nilpotent commutative generators. In this approach the motion groups of constant curvature spaces (or Cayley–Klein groups) are realized as matrix groups of special form over $P_n(\iota)$ and can be obtained from the simple classical orthogonal group by substitution of its matrix elements for Pimenov algebra elements. It turns out that such substitution coincides with the introduction of Wigner–Inönü contraction parameter ϵ. So our approach demonstrates that the existence of the corresponding structures over algebra $P_n(\iota)$ is the mathematical base of the contraction method.

It should be noted that both approaches supplement each other and in the final analysis give the same results. Nilpotent generators are more suitable in the mathematical consideration of contractions whereas the contraction parameter continuously tending to zero corresponds more to the physical intuition where a physical system continuously changes its state and smoothly goes into its limit state. In chapter 5, devoted to the applications of the contraction method to the modern theory of the elementary particles, the traditional approach is used.

It is well known in geometry (see, for example, review [Yaglom, Rosenfeld and Yasinskaya (1964)]) that there are 3^n different geometries of dimension n, which admit the motion group of maximal order. R.I. Pimenov suggested [Pimenov (1959, 1965)] a unified axiomatic description of all 3^n geometries of constant curvature (or Cayley–Klein geometries) and demonstrated that all these geometries can be locally simulated in some region of n-dimension spherical space with named coordinates, which can be real, imaginary and nilpotent ones. According to Erlanger program by F. Klein the main content in geometry is its motion group whereas the properties of transforming objects are secondary. The motion group of n-dimensional spherical space is isomorphic to the orthogonal group $SO(n+1)$. The groups obtained from $SO(n+1)$ by contractions and analytical continuations are isomorphic to the motion groups of Cayley–Klein spaces. This correspondence provides the geometrical interpretation of Cayley–Klein contraction scheme. By analogy this interpretation is transferred to the contractions of other algebraic structures.

The aim of this book is to develop contraction method for Cayley–Klein orthogonal and unitary groups (algebras) and apply it to the investigation of physical structures. The contraction method apart from being of interest to group theory itself is of interest to theoretical physics too. If there is a group-theoretical description of a physical system, then the contraction of its symmetry group corresponds to some limit case of the system under consideration. So the reformulation of the system description in terms of this method and the subsequent physical interpretations of contraction parameters give an opportunity to study different limit behaviours of the physical system. Examples of such an approach are given in chapter 2 for the space–time models, in chapter 3 for the Jordan–Schwinger representations of groups which are closely connected to the properties of stationary quantum systems whose Hamiltonians are quadratic in creation and annihilation operators. In chapter 5 the modified Standard Model with contracted gauge group is considered. The contraction parameter tending to zero is associated with the inverse average energy (temperature) of the Universe which makes it possible to re-establish the evolution of particles and their interactions in the early Universe.

It is likely that the developed formalism is essential in constructing "general theory of physical systems" where it will necessarily turn the group-invariant study of a single physical theory in Klein's understanding (i.e. characterized by symmetry group) to a simultaneous study of a set of limit theories. Then some physical and geometrical properties will be the invariant properties of all sets of theories and they should be considered in the first place. Other properties will be relevant only for particular representatives and will change under limit transition from one theory to another [Zaitsev (1974)].

In chapter 1, definitions of Cayley–Klein orthogonal and unitary groups are given; their generators, commutators and Casimir operators are obtained by transformations of the corresponding quantities of classical groups.

In chapter 4, contractions of irreducible representations of unitary and orthogonal algebras in Gel'fand–Tsetlin basis [Gromov (1991, 1992)], which are especially convenient for applications in quantum physics, are considered. Possible contractions, which give different representations of contracted algebras, are found and general contractions leading to representations with nonzero eigenvalues of Casimir operators are studied in detail. For multi-parametric contractions, when a contracted algebra is a semidirect sum of a nilpotent radical and a semisimple subalgebra, our method ensures the construction of irreducible representations for such algebras starting from well known irreducible representations of classical algebras. When algebras contracted on different parameters are isomorphic, the irreducible representations in different bases (discrete and continuous) are obtained.

The list of references does not pretend to be complete and reflects the author's interests.

N. A. Gromov

Contents

Chapter 1

The Cayley–Klein groups and algebras

In this chapter Pimenov algebra with commutative nilpotent generators is introduced. The definitions of orthogonal, unitary and symplectic Cayley–Klein groups are given. It is shown that the basic algebraic constructions, characterizing Cayley–Klein groups, can be found using simple transformations from the corresponding constructions for classical groups. The theorem on the classifications of transitions is proved, which shows that all Cayley–Klein groups can be obtained not only from simple classical groups. As a starting point, one can choose any pseudogroup as well.

1.1 Dual numbers and the Pimenov algebra

1.1.1 *Dual numbers*

Dual numbers were introduced by Clifford (1873) as far back as in the nineteenth century. They were used by Kotel'nikov (1895) for constructing his theory of screws in three-dimensional spaces of Euclid, Lobachevsky and Riemann; by Rosenfeld (1955) and jointly with Karpova (1966) for description of non-Euclidean spaces; by Pimenov (1959, 1965) for axiomatic study of spaces with a constant curvature. Some applications of dual numbers in kinematics can be found in the work by Yaglom (1979). The applications of dual numbers in geometry and in theory of group representations were discussed by Kisil (2012a). Fine distinctions between quantum and classical mechanics were investigated with the help of dual numbers [Kisil (2012b,c)]. The theory of dual numbers as number systems is exposed in monographs by Zeiliger (1934) and Blokh (1982). Nevertheless, it is impossible to say that dual numbers are well known, so we start with their description.

Definition 1.1. By the associative algebra of rank n over the real numbers field $\mathbb{R}$ we mean n-dimensional vector space over this field, on which the operation of multiplication is defined, associative $a(bc) = (ab)c$, distributive with respect to addition $(a+b)c = ac + bc$ and related to the multiplication of elements by real numbers as follows $(ka)b = k(ab) = a(kb)$, where a, b, c are the elements of algebra; k is a real number. If there exists an element e of algebra, such that the relations $ae = ea = a$ are valid for any element a of algebra, then the element e is called a unit.

Definition 1.2. Dual numbers $a = a_0 e_0 + a_1 e_1$, $a_0, a_1 \in \mathbb{R}$ are the elements of associative algebra of rank 2 with the unit and the generators satisfying the following conditions: $e_0^2 = e_0$, $e_0 e_1 = e_1 e_0$, $e_1^2 = 0$.

This associative algebra is commutative and e_0 is its unit. Therefore, we shall write 1 instead of e_0 and denote generator e_1 by ι_1 (the Greek letter "*iota*") and call it a (purely) dual unit.

For a sum, a product and a quotient of dual numbers a and b, we have

$$a + b = (a_0 + \iota_1 a_1) + (b_0 + \iota_1 b_1) = a_0 + b_0 + \iota_1(a_1 + b_1),$$

$$ab = (a_0 + \iota_1 a_1)(b_0 + \iota_1 b_1) = a_0 b_0 + \iota_1(a_1 b_0 + a_0 b_1), \qquad (1.1)$$

$$\frac{a}{b} = \frac{a_0 + \iota_1 a_1}{b_0 + \iota_1 b_1} = \frac{a_0}{b_0} + \iota_1 \left(\frac{a_1}{b_0} - a_0 \frac{b_1}{b_0^2} \right).$$

Division cannot always be carried out. Purely dual numbers $a_1 \iota_1$ do not have an inverse element. Therefore dual numbers do not form a number field. As an algebraic structure they perform a ring. Dual numbers satisfy the equality, $a = b$, if their real parts and purely dual parts satisfy, respectively, $a_0 = b_0$ and $a_1 = b_1$. Thus, the equation $a_1 \iota_1 = b_1 \iota_1$ has the unique solution $a_1 = b_1$ for $a_1, b_1 \neq 0$. This fact can be written formally as $\iota_1 / \iota_1 = 1$ and this is how the last relation has to be interpreted because division $1/\iota_1$ is not defined.

Functions of dual variable $x = x_0 + \iota_1 x_1$ are defined by their Taylor expansion

$$f(x) = f(x_0) + \iota_1 x_1 \frac{\partial f(x_0)}{\partial x_0}, \qquad (1.2)$$

where all terms with coefficients $\iota_1^2, \iota_1^3, \ldots$ are omitted. In particular, for dual x we have

$$\sin x = \sin x_0 + \iota_1 x_1 \cos x_0, \quad \sin(\iota_1 x_1) = \iota_1 x_1,$$
$$\cos x = \cos x_0 - \iota_1 x_1 \sin x_0, \quad \cos(\iota_1 x_1) = 1. \qquad (1.3)$$

According to (1.2), the difference of two functions of dual variable can be presented as

$$f(x) - h(x) = f(x_0) - h(x_0) + \iota_1 x_1 \left(\frac{\partial f(x_0)}{\partial x_0} - \frac{\partial h(x_0)}{\partial x_0} \right), \qquad (1.4)$$

therefore, if real parts $f(x_0)$ and $h(x_0)$ of functions coincide, then functions $f(x)$ and $h(x)$ also coincide. Using this fact, it was shown by Zeiliger (1934) that in the domain of dual numbers all identities of algebra and trigonometry, and all theorems of differential and integral calculus remain valid. In particular, the derivative of a function of a dual variable over a dual variable can be found as

$$\frac{df(x)}{dx} = \frac{\partial f(x_0)}{\partial x_0} + \iota_1 x_1 \left(\frac{\partial^2 f(x_0)}{\partial x_0^2} \right). \qquad (1.5)$$

1.1.2 The Pimenov algebra

Let us now consider a more general situation, where several nilpotent units are taken as generators of an associative algebra with a unit. (Later, we will use the name *nilpotent unit* as well as *dual unit*). R.I. Pimenov was the first who introduced [Pimenov (1959, 1965)] several nilpotent commutative units and used them for the unified axiomatic description of spaces with constant curvature. Named in his honor, we introduce the Pimenov algebra and denote it as $\mathbf{P}_n(\iota)$.

Definition 1.3. Pimenov algebra $\mathbf{P}_n(\iota)$ is an associative algebra with a unit and n nilpotent generators $\iota_1, \iota_2, \ldots, \iota_n$ with properties: $\iota_k \iota_p = \iota_p \iota_k \neq 0$, $k \neq p$, $\iota_k^2 = 0$, $p, k = 1, 2, \ldots, n$.

Any element of algebra $\mathbf{P}_n(\iota)$ is a linear combination of monomials $\iota_{k_1} \iota_{k_2} \cdots \iota_{k_r}$, $k_1 < k_2 < \cdots < k_r$, which together with a unit element make a basis in algebra as in a linear space of dimension 2^n:

$$a = a_0 + \sum_{r=1}^{n} \sum_{k_1, \ldots, k_r = 1}^{n} a_{k_1 \ldots k_r} \iota_{k_1} \cdots \iota_{k_r}. \qquad (1.6)$$

This notation becomes unique, if we put an additional requirement $k_1 < k_2 < \cdots < k_r$ or condition of symmetry of coefficients $a_{k_1 \ldots k_r}$ with respect to indices $k_1, \ldots, k_r$. Two elements a, b of algebra $\mathbf{P}_n(\iota)$ coincide, if their coefficients in the expansion (1.6) are equal, i.e. $a_0 = b_0$, $a_{k_1 \ldots k_r} = b_{k_1 \ldots k_r}$. As in the case of dual numbers, this definition of equality of the elements of algebra $\mathbf{P}_n(\iota)$ is equivalent to the possibility of reducing the same (with the same index) nilpotent units $\iota_k/\iota_k = 1$, $k = 1, 2, \ldots, n$ (but not ι_k/ι_m or ι_m/ι_k, $k \neq m$, as far as such expressions are undefined).

Here it is appropriate to compare Pimenov algebra $\mathbf{P}_n(\iota)$ with Grassmann algebra $\Gamma_{2n}(\epsilon)$, i.e. associative algebra with a unit, where a set of nilpotent generators $\epsilon_1, \epsilon_2, \ldots, \epsilon_{2n}$, $\epsilon_k^2 = 0$ exhibits the properties of anti-commutativity $\epsilon_k \epsilon_p = -\epsilon_p \epsilon_k \neq 0$, $p \neq k$, $p, k = 1, \ldots, 2n$. Any element f of Grassmann algebra $\Gamma_{2n}(\epsilon)$ can be expressed [Beresin (1987)] as

$$f(\epsilon) = f(0) + \sum_{r=1}^{2n} \sum_{k_1,\ldots,k_r=1}^{2n} f_{k_1 \cdots k_r} \epsilon_{k_1} \cdots \epsilon_{k_1}. \tag{1.7}$$

The representation is unique, if one requires $k_1 < k_2 < \cdots < k_r$ or applies the condition of skew-symmetry $f_{k_1 \cdots k_r}$ with respect to indices $k_1, \ldots, k_r$. If only the terms with an even r in the expansion (1.7) differ from zero, then the element f is even with respect to the set of canonical generators ϵ_k. Likewise, if only the terms with an odd r differ from zero, then f is called an odd element. As a linear space, Grassmann algebra splits into even Γ_{2n}^0 and odd Γ_{2n}' subspaces: $\Gamma_{2n}(\epsilon) = \Gamma_{2n}^0 + \Gamma_{2n}^1$, where Γ_{2n}^0 is not only a subspace, but also a subalgebra.

Let us consider nonzero products $\epsilon_{2k-1} \epsilon_{2k}$, $k = 1, 2, \ldots, n$ of the generators of Grassmann algebra $\Gamma_{2n}(\epsilon)$. It is easy to see that these products possess the same properties as generators $\iota_k = \epsilon_{2k-1} \epsilon_{2k}$, $k = 1, 2, \ldots, n$. Thus Pimenov algebra $\mathbf{P}_n(\iota)$ is a subalgebra of the even part Γ_{2n}^0 of Grassmann algebra $\Gamma_{2n}(\epsilon)$. It is worth mentioning that the even products of Grassmannian anticommuting generators are also called para-Grassmannian variables. The latter are employed for classical and quantum descriptions of massive and massless particles with an integer spin, [Gershun and Tkach (1984, 1985); Duplii (1988)] and in theory of strings [Zheltuhin (1985)].

1.2   The Cayley–Klein orthogonal groups and algebras

1.2.1   *Three fundamental geometries on a line*

Let us introduce elliptic geometry on a line. Let us consider a circle $\mathbf{S}_1^* = \{x_0^{*2} + x_1^{*2} = 1\}$ on the Euclid plane $\mathbf{R}_2$. The rotations $x'^* = g(\varphi^*) x^*$, i.e.

$$\begin{aligned} x_0^{*'} &= x_0^* \cos \varphi^* - x_1^* \sin \varphi^*, \\ x_1^{*'} &= x_0^* \sin \varphi^* + x_1^* \cos \varphi^* \end{aligned} \tag{1.8}$$

of group $SO(2)$ bring the circle into itself. Let us identify diametrically the opposite points of the circle and introduce an internal coordinate $w^* = x_1^*/x_0^*$. Then the following transformations correspond to the rotations (1.8) in $\mathbf{R}_2$ for $\varphi^* \in (-\pi/2, \pi/2)$:

$$w^{*'} = \frac{w^* - a^*}{1 + w^*a^*}, \quad a^* = \tan\varphi^*, \quad a^* \in \mathbb{R}. \tag{1.9}$$

These transformations make a group of translations (motions) G_1 of an elliptic line with the rule of composition

$$a^{*'} = \frac{a^* + a_1^*}{1 - a^*a_1^*}. \tag{1.10}$$

Let us consider the representation of the group $SO(2)$ in the space of differentiable functions on $\mathbf{R}_2$, defined by the left shifts

$$T(g(\varphi^*))f(x^*) = f(g^{-1}(\varphi^*)x^*). \tag{1.11}$$

The generator of the representation

$$X^*f(x^*) = \frac{d(T(g(\varphi^*))f(x^*))}{d\varphi^*}\Big|_{\varphi^*=0}, \tag{1.12}$$

corresponding to the transformation (1.8), can be easily found:

$$X^*(x_0^*, x_1^*) = x_1^*\frac{\partial}{\partial x_0^*} - x_0^*\frac{\partial}{\partial x_1^*}. \tag{1.13}$$

For the representation of group G_1 — defined by the left shifts in the space of differentiable functions on an elliptic line — the generator Z^*, corresponding to the transformation (1.9), can be written as

$$Z^*(w^*) = (1 + w^{*2})\frac{\partial}{\partial w^*}. \tag{1.14}$$

It is worth mentioning that the matrix generator

$$Y^* = \begin{pmatrix} 0 & -1 \\ 1 & 0 \end{pmatrix} \tag{1.15}$$

corresponds to rotations $g(\varphi^*) \in SO(2)$.

The transformation of Euclidean plane $\mathbf{R}_2$, consisting of multiplication of Cartesian coordinate x_1 by parameter j_1, namely

$$\phi : \mathbf{R}_2 \to \mathbf{R}_2(j_1)$$
$$\phi x_0^* = x_0, \quad \phi x_1^* = j_1 x_1, \tag{1.16}$$

where $j_1 = 1, \iota_1, i$, brings $\mathbf{R}_2$ into plane $\mathbf{R}_2(j_1)$; the geometry of the latter is defined by metrics $\mathbf{x}^2(j_1) = x_0^2 + j_1^2 x_1^2$. It is easy to see that $\mathbf{R}_2(j_1 = i)$ is a Minkowski plane and $\mathbf{R}_2(j_1 = \iota_1)$ is a Galilean plane.

Our main idea is that the transformation of geometries (1.16) induces the transformation of the corresponding motion groups and their algebras. Let us show how to derive these transformations.

The definition of angle measure in Euclidean plane R_2 is determined by the ratio x_1^*/x_0^*, which under the transformation (1.16) turns into $j_1 x_1/x_0$, i.e. angles are transformed according to the rule $\phi\varphi^* = j_1\varphi$. The asterisk marks the initial quantities (coordinates, angles, generators and so on). The transformed quantities are denoted by the same symbols without asterisk. Changing the coordinates in (1.8) according to (1.16) and the angles according to the derived transformation rule, and multiplying both sides of the second equation by j_1^{-1}, we get the rotations in the plane $\mathbf{R}_2(j_1)$:

$$
\begin{aligned}
x_0' &= x_0 \cos j_1\varphi - x_1 j_1 \sin j_1\varphi, \\
x_1' &= x_0 \frac{1}{j_1} \sin j_1\varphi + x_1 \cos j_1\varphi,
\end{aligned}
\tag{1.17}
$$

which make the group $SO(2; j_1)$. According to (1.3), $\cos \iota_1\varphi = 1$, $\sin \iota_1\varphi = \iota_1\varphi$, therefore the transformations of group $SO(2; \iota_1)$ are Galilean transformations and the elements of group $SO(2; i)$ are Lorentz transformations, if x_0 is interpreted as time, and x_1 — as a spatial coordinate. The domain of definition $\Phi(j_1)$ of the group parameter φ is $\Phi(1) = (-\pi/2, \pi/2)$, $\Phi(\iota_1) = \Phi(i) = \mathbb{R}$.

The rotations (1.17) preserve the circle $\mathbf{S}_1(j_1) = \{x_0^2 + j_1^2 x_1^2 = 1\}$ (Fig. 1.1) in the plane $\mathbf{R}_2(j_1)$. The identification of diametrically opposite points gives the upper semicircle (for $j_1 = 1$) and the connected component of the sphere (circle), passing through the point $(x_0 = 1, x_1 = 0)$, for $j_1 = \iota_1, i$. The internal coordinate on the circle w^* is transformed according to the rule $\phi w^* = j_1 w$. Substituting in (1.9) and canceling j_1 out of both sides we get the formula for translations on a line:

$$
w' = \frac{w - a}{1 + j_1^2 wa}, \quad a = \frac{1}{j_1} \tan j_1\varphi \in \mathbb{R},
\tag{1.18}
$$

which make group $G_1(j_1)$, i.e. the group of motions of the elliptic line $\mathbf{S}_1(1)$ for $j_1 = 1$, the parabolic line $\mathbf{S}_1(\iota_1)$ for $j_1 = \iota_1$, and the hyperbolic line $\mathbf{S}_1(i)$ for $j_1 = i$.

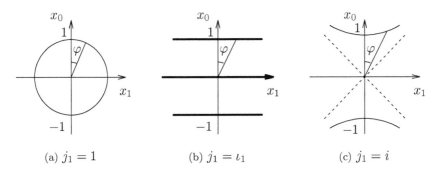

(a) $j_1 = 1$ (b) $j_1 = \iota_1$ (c) $j_1 = i$

Fig. 1.1   The circles of unit radius on the planes $\mathbf{R}_2(j_1)$.

In the space of differentiable functions on $\mathbf{R}_2(j_1)$ the generator $X(x)$ of the representation of group $SO(2; j_1)$ is defined by the relation (1.12), where all quantities are taken without asterisks. Under the transformation (1.16) derivative $d/d\varphi^*$ turns $j_1^{-1}(d/d\varphi)$; therefore, to obtain derivative $d/d\varphi$ the generator X^* must be multiplied by j_1, i.e. the generators $X^*(\phi x)$ and $X(x)$ are interrelated by the transformation

$$X(x) = j_1 X^*(\phi x^*) = j_1^2 x_1 \frac{\partial}{\partial x_0} - x_0 \frac{\partial}{\partial x_1}. \tag{1.19}$$

The generator Z is transformed according to the same rule:

$$Z(w) = j_1 Z^*(\phi w^*) = (1 + j_1^2 w^2) \frac{\partial}{\partial w}. \tag{1.20}$$

The transformation rule for the matrix generator of the rotation Y is as follows:

$$Y = j_1 Y^*(\rightarrow) = j_1 \begin{pmatrix} 0 & -j_1 \\ j_1^{-1} & 0 \end{pmatrix} = \begin{pmatrix} 0 & -j_1^2 \\ 1 & 0 \end{pmatrix}. \tag{1.21}$$

Expressions (1.17)–(1.21) describe Cayley–Klein space and group in the traditional way with the help of real coordinates, generators and so on. Such an approach was used in [Gromov (1990)] and will be used in chapter 3 for Jordan–Schwinger representations of Cayley–Klein groups. There is another way of describing Cayley–Klein spaces with the help of the named (i.e. real, nilpotent or imaginary) coordinates of the form $j_1 x_1$, when both sides of the second equation are not multiplied by j_1^{-1} under transformation (1.16) and the substitution $\phi\varphi^* = j_1\varphi$ in (1.8). Then the rotations on the plane $\mathbf{R}_2(j_1)$ with coordinates x_0, j_1x_1 are written in the form

$$\begin{pmatrix} x_0' \\ j_1x_1' \end{pmatrix} = \begin{pmatrix} \cos j_1\varphi & -\sin j_1\varphi \\ \sin j_1\varphi & \cos j_1\varphi \end{pmatrix} \begin{pmatrix} x_0 \\ j_1x_1 \end{pmatrix}. \tag{1.22}$$

These rotations form group $SO(2; j_1)$, whose matrix generator is as follows

$$Y = j_1Y^* = \begin{pmatrix} 0 & -j_1 \\ j_1 & 0 \end{pmatrix}. \tag{1.23}$$

The symbol Y^* instead of $Y^*(\rightarrow)$ in (1.21) means that the generator Y^* (1.15) is not transformed. It is the second approach that we shall use in this book. One of its advantages is that for $j_1 = \iota_1$ the rotation matrix (1.22) from group $SO(2; \iota_1)$

$$\begin{pmatrix} 1 & -\iota_1\varphi \\ \iota_1\varphi & 1 \end{pmatrix}, \tag{1.24}$$

depend on group parameter φ, whereas for $j_1 \rightarrow 0$ it is equal to the unit matrix.

The group of motions $G_1(j_1)$ of one-dimensional Cayley–Klein space $\mathbf{S}_1(j_1)$ is closely connected with rotation group $SO(2; j_1)$ in space $\mathbf{R}_2(j_1)$. Therefore, by Cayley–Klein space we shall implicitly mean both $\mathbf{S}_1(j_1)$ and $\mathbf{R}_2(j_1)$, and by their groups of motion — both $G_1(j_1)$ and $SO(2; j_1)$. We shall follow the same rule in the case of higher dimensions.

We have studied comprehensively the simplest case of groups $SO(2; j_1)$, $G_1(j_1)$ because the main ideas on methods of transitions reveal themselves in the clearest way, free from convoluted calculations. These ideas are as follows: (a) to define the transformation (1.16) from Euclidean space to arbitrary Cayley–Klein space; (b) to find the rules of transformations of motion, generators, etc. of the group; (c) using the approach exposed in (b) to derive motion, generators, etc. of Cayley–Klein group from the corresponding quantities of classical orthogonal group. The method of transitions, in spite of its simplicity, enables us to describe all Cayley–Klein groups with information of only classical orthogonal ones.

1.2.2   *Nine Cayley–Klein groups*

Mapping

$$\phi : \mathbf{R}_3 \rightarrow \mathbf{R}_3(j)$$
$$\phi x_0^* = x_0, \quad \phi x_1^* = j_1x_1, \quad \phi x_2^* = j_1j_2x_2, \tag{1.25}$$

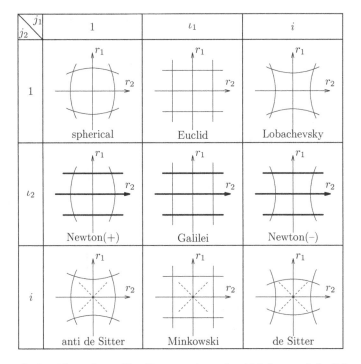

Fig. 1.2 Cayley–Klein planes. The fibers are shown by thick lines and the light cone in $(1+1)$ kinematics are shown by dashed lines. Internal Beltrami coordinates take values $r_1 = x_1/x_0$, $r_2 = x_2/x_0$.

where $j = (j_1, j_2)$, $j_1 = 1, \iota_1, i$, $j_2 = 1, \iota_2, i$, turns three-dimensional Euclidean space into spaces $\mathbf{R}_3(j)$. The nine geometries of Cayley–Klein planes are realized on the spheres (or connected components of spheres)

$$\mathbf{S}_2(j) = \{x_0^2 + j_1^2 x_1^2 + j_1^2 j_2^2 x_2^2 = 1\} \qquad (1.26)$$

in these spaces. The interrelation of the geometries and values of parameters j is clear from Fig. 1.2.

Rotation angle φ_{rs} in the coordinate plane $\{x_r, x_s\}$, $r < s$, $r, s = 0, 1, 2$, is determined by the ratio x_s/x_r and under the mapping (1.25) is transformed as $\varphi_{rs}^* \to \varphi_{rs}(r, s)$, where

$$(i, k) = \prod_{l=\min(i,k)+1}^{\max i,k} j_l, \quad (k, k) = 1. \qquad (1.27)$$

Therefore for one-parametric rotations in the plane $\{x_r, x_s\}$ of space $\mathbf{R}_3(j)$ the following relations are valid:

$$
\begin{aligned}
(0,r)x'_r &= x_r(0,r)\cos(\varphi_{rs}(r,s)) - x_s(0,s)\sin(\varphi_{rs}(r,s)), \\
(0,s)x'_s &= x_r(0,r)\sin(\varphi_{rs}(r,s)) + x_s(0,s)\cos(\varphi_{rs}(r,s)).
\end{aligned}
\tag{1.28}
$$

The rest of the coordinates are not changed: $x'_p = x_p$, $p \neq r, s$.

It is easy to find the matrix generators of the rotations (1.28)

$$
Y_{01} = j_1 Y_{01}^* = \begin{pmatrix} 0 & -j_1 & 0 \\ j_1 & 0 & 0 \\ 0 & 0 & 0 \end{pmatrix}, \quad
Y_{12} = j_2 Y_{12}^* = \begin{pmatrix} 0 & 0 & 0 \\ 0 & 0 & -j_2 \\ 0 & j_2 & 0 \end{pmatrix},
$$

$$
\tag{1.29}
$$

$$
Y_{02} = j_1 j_2 Y_{02}^* = \begin{pmatrix} 0 & 0 & -j_1 j_2 \\ 0 & 0 & 0 \\ j_1 j_2 & 0 & 0 \end{pmatrix}.
$$

They make a basis of Lie algebra $so(3; j)$. The rule of transformations for the generators of representation of group $SO(3; j)$ in the space of differentiable functions on $\mathbf{R}_3(j)$ by left shifts coincides with the rule of transformations for parameters φ_{rs} and can be written as follows [Gromov (1981)]:

$$
X_{rs}(x) = (r,s)X_{rs}^*(\phi x^*),
\tag{1.30}
$$

and the generators themselves as

$$
X_{rs}(x) = (r,s)^2 x_s \frac{\partial}{\partial x_r} - x_r \frac{\partial}{\partial x_s}.
\tag{1.31}
$$

Knowing the generators, one can evaluate their commutators. But we shall derive the commutators from the commutation relations of algebra $so(3)$. Let us introduce new notations for the generators $X_{01}^* = H^*$, $X_{02}^* = P^*$, $X_{12}^* = K^*$. As it is well-known, the commutators of Lie algebra $so(3)$ can be written as follows:

$$
[H^*, P^*] = K^*, \quad [P^*, K^*] = H^*, \quad [H^*, K^*] = -P^*.
\tag{1.32}
$$

Generators of algebra $so(3)$ are transformed according to the rule $H = j_1 H^*$, $P = j_1 j_2 P^*$, $K = j_2 K^*$, i.e. $H^* = j_1^{-1} H$, $P^* = j_1^{-1} j_2^{-1} P$,

$K^* = j_2^{-1}K$. Substituting these expressions in (1.32) and multiplying each commutator by a factor, equal to the denominator on the left side of each equation, i.e. the first — by $j_1^2 j_2$, the second — by $j_1 j_2^2$, and the third — by $j_1 j_2$, we get commutators of Lie algebra for group $SO(3; j)$:

$$[H, P] = j_1^2 K, \quad [P, K] = j_2^2 H, \quad [H, K] = -P. \tag{1.33}$$

Cayley–Klein spaces $\mathbf{S}_2(j)$ (or spaces of constant curvature) for $j_1 = 1$, ι_1, i, $j_2 = \iota_2, i$ can serve as models of kinematics, i.e. space-time geometries. In this case, internal coordinate $t = x_1/x_0$ can be interpreted as the temporal axis, and internal coordinate $r = x_2/x_0$ as the spatial one. Then H is the generator of the temporal shift, P is the generator of the spatial shift, and K is the generator of Galilean transformation for $j_2 = \iota_2$ or Lorentz transformation for $j_2 = i$. The semispherical group $SO(3; 1, \iota_2)$ (or Newton group) is isomorphic to the cylindrical group, which describes movement of a point on a cylindrical surface. This group is interpreted as the $E(2)$-like little group for massless particles [Kim and Wigner (1987)].

The final relations should not involve division by a nilpotent number. This requirement suggests the way of finding the rule of transformations for algebraic constructions. Let an algebraic quantity $Q^* = Q^*(A_1^*, \ldots, A_k^*)$ be expressed in terms of quantities $A_1^*, \ldots, A_k^*$ with a known rule of transformation under mapping ϕ, for example, $A_1 = J_1 A_1^*, \ldots, A_k = J_k A_k^*$, where coefficients $J_1, \ldots, J_k$ are some products of parameters j. Substituting $A_1^* = J_1^{-1} A_1, \ldots, A_k^* = J_k^{-1} A_k$ in the relation for Q^*, we get the formula $Q^*(J_1^{-1} A_1, \ldots, J_k^{-1} A_k)$, involving, in general, indeterminate expressions, when parameters j are equal to the nilpotent units. For this reason the last formula should be multiplied by minimal coefficient J such that the final formula would not involve indeterminate expressions:

$$Q = JQ^*(J_1^{-1} A_1, \ldots, J_k^{-1} A_k). \tag{1.34}$$

Then (1.34) is the rule of transformation for quantity Q under mapping ϕ.

Such method, stemmed out directly from the definition of coinciding elements of Pimenov algebra $\mathbf{P}_n(\iota)$, turns out to be very useful and widely employed. The rule of transformation (1.34) for algebraic quantity Q, derived from the requirement of absence of indeterminate expressions for nilpotent values of parameters j, is automatically satisfied for imaginary values of these parameters.

Let us exemplify this rule by Casimir operator. The only Casimir operator for algebra $so(3)$ is

$$C_2^*(H^*, \ldots) = H^{*2} + P^{*2} + K^{*2}. \tag{1.35}$$

Substituting $H^* = j_1^{-1}H$, $P^* = j_1^{-1}j_2^{-1}P$, $K^* = j_2^{-1}K$ in (1.35), we get

$$C_2^*(j_1^{-1}H, \ldots) = j_1^{-2}H^2 + j_1^{-2}j_2^{-2}P^2 + j_2^{-2}K^2. \tag{1.36}$$

The most singular factor for $j_1 = \iota_1$, $j_2 = \iota_2$ is coefficient $(j_1 j_2)^{-2}$ of the term P^2. Multiplying both sides of the equation (1.36) by $(j_1 j_2)^2$, we get rid of the indeterminate expressions and derive the rule of transformation and Casimir operator for algebra $so(3; j)$:

$$C_2(j; H, \ldots) = j_1^2 j_2^2 C_2^*(j_1^{-1}H, \ldots) = j_2^2 H^2 + P^2 + j_1^2 K^2. \tag{1.37}$$

As it is known, Casimir operator for two dimensional Galilean algebra $so(3; \iota_1, \iota_2)$ is $C_2(\iota_1, \iota_2) = P^2$ (see [Levy-Leblond (1965)]), for Poincaré algebra $so(3; \iota_1, i)$ is $C(\iota_1, i) = P^2 - H^2$, and for algebra $so(3; i; 1) = so(2, 1)$ is $C_2(i, 1) = H^2 + P^2 - K^2$ (see [Mukunda (1967)]). All these Casimir operators can be obtained from (1.37) for the corresponding values of parameters j.

The matrix generators (1.29) make the basis of fundamental representation of Lie algebra $so(3; j)$ of group $SO(3; j)$. Using exponential mapping one can relate the general element

$$Y(\mathbf{r}; j) = r_1 Y_{01} + r_2 Y_{02} + r_3 Y_{12} = \begin{pmatrix} 0 & -j_1 r_1 & -j_1 j_2 r_2 \\ j_1 r_1 & 0 & -j_2 r_3 \\ j_1 j_2 r_2 & j_2 r_3 & 0 \end{pmatrix} \tag{1.38}$$

of algebra $so(3; j)$ to the finite rotation $g(\mathbf{r}; j) = \exp Y(\mathbf{r}; j)$:

$$g(\mathbf{r}; j) = E \cos(r) + Y(\mathbf{r}; j)\frac{\sin r}{r} + Y'(\mathbf{r}, j)\frac{1 - \cos r}{r^2},$$

$$Y'(\mathbf{r}; j) = \begin{pmatrix} j_2^2 r_3^2 & -j_1 j_2^2 r_2 r_3 & j_1 j_2 r_1 r_3 \\ -j_1 j_2^2 r_2 r_3 & j_1^2 j_2^2 r_2^2 & -j_1^2 j_2 r_1 r_2 \\ j_1 j_2 r_1 r_3 & -j_1^2 j_2 r_1 r_2 & j_1^2 r_1^2 \end{pmatrix}, \tag{1.39}$$

$$r^2 = j_1^2 r_1^2 + j_1^2 j_2^2 r_2^2 + j_2^2 r_3^2,$$

acting on vector $(x_0, j_1 x_1, j_1 j_2 x_2)^t \in \mathbf{R}_3(j)$ with the named components.

The disadvantage of the general parametrization (1.38), (1.39) is the complexity of the composition rule for parameters **r** under group multiplication. F.I. Fedorov (1979) has proposed parametrization of rotation group $SO(3)$ for which the group composition law is especially simple. It turns out that it is possible to construct analogues of such parametrization for all groups $SO(3; j)$ [Gromov (1984)]. The matrix of the finite rotations of group $SO(3; j)$ can be written as follows

$$g(\mathbf{c}; j) = \frac{1 + c^*(j)}{1 - c^*(j)} = 1 + 2\frac{c^*(j) + c^{*2}(j)}{1 + c^2(j)},$$

$$c^*(j) = \begin{pmatrix} 0 & -j_1^2 c_3 & j_1^2 j_2^2 c_2 \\ c_3 & 0 & -j_2^2 c_1 \\ -c_2 & c_1 & 0 \end{pmatrix}, \qquad (1.40)$$

$$c^2(j) = j_2^2 c_1^2 + j_1^2 j_2^2 c_2^2 + j_1^2 c_3^2,$$

and parameters $\mathbf{c}''$ correspond to matrix $g(\mathbf{c}''; j) = g(\mathbf{c}; j)g(\mathbf{c}'; j)$. These parameters can be expressed in terms of **c** and $\mathbf{c}'$ as follows

$$\mathbf{c}'' = \frac{\mathbf{c} + \mathbf{c}' + [\mathbf{c}, \mathbf{c}']_j}{1 - (\mathbf{c}, \mathbf{c}')_j}. \qquad (1.41)$$

Here the scalar product of vectors **c** and $\mathbf{c}'$ is given by (1.40), and the vector product is given by

$$[\mathbf{c}, \mathbf{c}']_j = (j_1^2[\mathbf{c}, \mathbf{c}']_1, \ [\mathbf{c}, \mathbf{c}']_2, \ j_2^2[\mathbf{c}, \mathbf{c}']_3), \qquad (1.42)$$

where $[\mathbf{c}, \mathbf{c}']_k$ are components of usual vector product in $\mathbf{R}_3$.

E.P. Wigner and E. Inönü have introduced the operation of contraction (limit transition) of groups, algebras and their representations. Under this operation the generators of the initial group (algebra) undergo transformation, depending on a parameter ϵ, so that for $\epsilon \neq 0$ this transformation is nonsingular and for $\epsilon \to 0$ it becomes singular. If the limits of the transformed generators exist for $\epsilon \to 0$, then they are the generators of a new (contracted) group (algebra), nonisomorphic to the initial one. It is worth mentioning that the transformation (1.30) of the generators of algebra $so(3)$ for the nilpotent values of parameters j is Wigner–Inönü contraction. Really, $X_{rs}^*(\phi x^*)$ is the singularly transformed generator of initial algebra $so(3)$; the product (r, s) plays the role of parameter ϵ, tending to zero, and the resulted generators $X_{rs}(\mathbf{x})$ are the generators of the contracted algebra $so(3; j)$.

Comparing the rule of transformation for generators (1.30) and the expression (1.38) for a general element of algebra $so(3)$, we find that for the imaginary values of parameters j some of the real group parameters r_k become imaginary, i.e. they are analytically continued from the field of real numbers to the field of complex numbers. In this case, orthogonal group $SO(3)$ is transformed into pseudoorthogonal group $SO(p,q)$, $p+q=3$. When parameters j take nilpotent values, real group parameters r_k become elements of Pimenov algebra $\mathbf{P}(\iota)$ of the special form and we get the contraction of group $SO(3)$. Thus, from the point of view of the group transformation under mapping ϕ, both operations — analytical continuation of groups and contraction of groups different at first sight — have the same nature: the continuation of real group parameters to the complex numbers field or to Pimenov algebra $\mathbf{P}(\iota)$.

1.2.3 *Extension to higher dimensions*

Cayley–Klein geometries of the dimension n are realized on spheres

$$\mathbf{S}_n(j) = \left\{ (x,x) = x_0^2 + \sum_{k=1}^{n} (0,k)^2 x_k^2 = 1 \right\} \tag{1.43}$$

in the spaces $\mathbf{R}_{n+1}(j)$ resulting from Euclidean space $\mathbf{R}_{n+1}$ under mapping

$$\phi : \mathbf{R}_{n+1} \to \mathbf{R}_{n+1}(j)$$
$$\phi x_0^* = x_0, \quad \phi x_k^* = (0,k)x_k, \quad k = 1, 2, \ldots, n, \tag{1.44}$$

where $j = (j_1, \ldots, j_n)$, $j_k = 1, \iota_k, i$, $k = 1, 2, \ldots, n$. If all parameters are equal to one, $j_k = 1$, then ϕ is identical mapping; if all or some parameters are imaginary $j_k = i$ and the others are equal to 1, then we obtain pseudoeuclidean spaces of different signature. The space $\mathbf{R}_{n+1}(j)$ is called non-fiber, if none of the parameters $j_1, \ldots, j_n$ take nilpotent value.

Definition 1.4. The space $\mathbf{R}_{n+1}(j)$ is called $(k_1, k_2, \ldots, k_p)$-fiber space, if $1 \leq k_1 < k_2 < \cdots < k_p \leq n$ and $j_{k_1} = \iota_{k_1}, \ldots, j_{k_p} = \iota_{k_p}$ and other $j_k = 1, i$.

These fiberings can be characterized by a set of consecutively nested projections $pr_1, pr_2, \ldots, pr_p$, where the base for pr_1 is the subspace, spanned over the basis vector $\{x_0, x_1, \ldots, x_{k_1-1}\}$, and the fiber is the subspace, spanned over $\{x_{k_1}, x_{k_1+1}, \ldots, x_n\}$; for pr_2 the base is the subspace

$\{x_{k_1}, x_{k_1+1}, \ldots, x_{k_2-1}\}$, and the fiber is the subspace $\{x_{k_2}, x_{k_1+1}, \ldots, x_n\}$ and so on.

From the mathematical point of view the fibering in the space $\mathbf{R}_{n+1}(j)$ is trivial [Bourbaki (2007)], i.e. its global and local structures are the same. From the physical point of view the fibering gives an opportunity to model quantities of different physical units. For example, in Galilean space, which is realized on the sphere $\mathbf{S}_4(\iota_1, \iota_2, 1, 1)$, there are time $t = x_1$, $[t] =$ sec and space $\mathbf{R}_3 = \{x_2, x_3, x_4\}$, $[x_k] =$ cm, $k = 2, 3, 4$ variables.

Definition 1.5. Group $SO(n + 1; j)$ consists of all the transformations of the space $\mathbf{R}_{n+1}(j)$ with unit determinant, keeping invariant the quadratic form (1.43).

The totality of all possible values of parameters j gives 3^n different Cayley–Klein spaces $\mathbf{R}_{n+1}(j)$ and geometries $\mathbf{S}_n(j)$. It is customary to identify the spaces (and their group of motions), if their metrics have the same signature; for instance: space $\mathbf{R}_3(1, i)$ with metric $x_0^2 + x_1^2 - x_2^2$ and space $\mathbf{R}_3(i, i)$ with metric $x_0^2 - x_1^2 + x_2^2$. But we have fixed Cartesian coordinate axes in $\mathbf{R}_{n+1}(j)$ ascribing to them fixed numbers, and for this reason, spaces $\mathbf{R}_3(1, i)$ and $\mathbf{R}_3(i, i)$ (and, correspondingly groups $SO(3; 1, i)$ and $SO(3; i, i)$) are different. Groups $SO(3; 1, i) \equiv SO(2, 1)$ and $SO(3; i, 1) \equiv SO(1, 2)$ are also considered to be different. This was made for the convenience of applications of the developed method.

Really, the application of some mathematical formalism in a concrete science means first of all substantial interpretation of base mathematical constructions. For example, if we interpret in space $\mathbf{R}_4(i, 1, 1)$ with metric $x_0^2 - x_1^2 - x_2^2 - x_3^2$ the first Cartesian coordinate x_0 as the time axis and the other x_1, x_2, x_3 as the space axes, then we get relativistic kinematics (space-time model). In this example, the substantial interpretation of coordinates is the numbers of Cartesian coordinate axes: axis number one, axis number two, etc.

The rotations in the two-dimensional plane $\{x_r, x_s\}$, the rule of transformation for representation generators and the generators themselves are given, correspondingly, by (1.28), (1.30), (1.31), where $r, s = 0, 1, \ldots, n$, $r < s$. For the nonzero elements of the matrix generators of rotations, the following relations are valid: $(Y_{rs})_{sr} = - (Y_{rs})_{rs} = (r, s)$. The commutation relations for Lie algebra $so(n + 1; j)$ can be most simply derived from the commutators of algebra $so(n + 1)$, as it has been done in section 1.2.2.

The nonzero commutators are

$$
[X_{r_1 s_1}, X_{r_2 s_2}] = \begin{cases} (r_1, s_1)^2 X_{s_1 s_2}, & r_1 = r_2,\ s_1 < s_2, \\ (r_2, s_2)^2 X_{r_1 r_2}, & r_1 < r_2,\ s_1 = s_2, \\ -X_{r_1 s_2}, & r_1 < r_2 = s_1 < s_2. \end{cases} \tag{1.45}
$$

Algebra $so(n+1)$ has $[(n+1)/2]$ independent Casimir operators, where $[x]$ is the integer part of a number x. As it is known [Barut and Raczka (1977)], for even $n = 2k$ Casimir operators are given by

$$
\hat{C}^*_{2p}(X^*_{rs}) = \sum_{a_1,\dots,a_p=0}^{n} X^*_{a_1 a_2} X^*_{a_2 a_3} \cdots X^*_{a_{2p} a_1}, \tag{1.46}
$$

where $p = 1, 2, \dots, k$. For odd $n = 2k+1$ the operator

$$
C'^*_n(X^*_{rs}) = \sum_{a_1,\dots,a_n=0}^{n} \epsilon_{a_1 a_2 \cdots a_n} X^*_{a_1 a_2} X^*_{a_3 a_4} \cdots X^*_{a_n a_{n+1}}, \tag{1.47}
$$

where $\epsilon_{a_1 \cdots a_n}$ is a completely antisymmetric unit tensor, must be added to the operators (1.46).

Casimir operators C^*_{2p} can be defined in another way [Gel'fand (1950)] as a sum of principal minors of order $2p$ for antisymmetric matrix A, composed of generators X^*_{rs}, i.e. $(A)_{rs} = X^*_{rs}, (A)_{sr} = -X^*_{rs}$. To obtain Casimir operators of algebra $so(n+1; j)$ we use the method in section 1.2.2. We find $X^*_{rs} = (r, s)^{-1} X_{rs}$ from (1.30) and substitute in (1.46). The most singular coefficient $(0, n)^{-2p}$ is that of the term $X_{0n} X_{n0} \cdots X_{n0}$ in (1.46). To eliminate it in a minimal manner we multiply $\hat{C}^*_{2p}$ by $(0, n)^{2p}$. Thus, the rule of transformation for Casimir operators $\hat{C}_{2p}$ is

$$
\hat{C}_{2p}(j; X_{rs}) = (0, n)^{2p} \hat{C}^*_{2p}((r, s)^{-1} X_{rs}), \tag{1.48}
$$

and Casimir operators themselves turn out to be

$$
\hat{C}_{2p}(j) = \sum_{a_1,\dots,a_{2p}=0}^{n} (0, n)^{2p} \prod_{v=1}^{2p} (r_v, s_v)^{-1} X_{a_1 a_2} \cdots X_{a_{2p} a_1}, \tag{1.49}
$$

where $r_v = \min(a_v, a_{v+1})$, $s_v = \max(a_v, a_{v+1})$, $v = 1, 2, \dots, 2p-1$, $r_{2p} = \min(a_1, a_{2p})$, $s_{2p} = \max(a_1, a_{2p})$.

For operators C_{2p} and C'_n the expression without singular terms can be obtained by multiplying them to the factor q — the least common denominator of coefficients of terms — which arises after the substitution of generators X for X^*. This least common denominator can be found by induction [Gromov (1981)]. We restrict ourselves to the final expressions for the rule of transformations for these Casimir operators:

$$
C_{2p}(j; X_{rs}) = \left(\prod_{m=1}^{p-1} j_m^{2m} j_{n-m+1}^{2m} \prod_{l=p}^{n-p+1} j_l^{2p} \right) C_{2p}^*(X_{rs}(r,s)^{-1}),
$$

$$
p = 1, 2, \ldots, k, \tag{1.50}
$$

$$
C_n'(j; X_{rs}) = \left(j_{(n+1)/2}^{(n+1)/2} \prod_{m=1}^{(n-1)/2} j_m^m j_{n-m+1}^m \right) C_n'^*(X_{rs}(r,s)^{-1}).
$$

Operator $C_{2p}(j)$ (or $C'(j)$) commutes with all generators X_{rs} of algebra $so(n+1; j)$. Really, evaluating zero commutator $[C_{2p}^*, X_{rs}^*]$, we get the same terms with the opposite signs. Under the transformations (1.30) and (1.48), both terms are multiplied by the same combination of parameters, which is a product of even powers of parameters. Therefore, both terms may change or retain their sign, or vanish, but in all cases their sum is equal to zero. Moreover, operators $C_{2p}(j)$ for $p = 1, 2, \ldots, k$ are linearly independent because they consist of different powers of generators X_{rs}.

The next question to be cleared up is as follows: do $[(n+1)/2]$ Casimir operators (1.50) exhaust all the invariant operators of algebra $so(n+1; j)$? The answer is given by the following theorem.

Theorem 1.1. *For any set of values of parameters j the number of invariant operators of algebra $so(n+1; j)$ is $[(n+1)/2]$.*

The proof is given in [Gromov (1990)]. Thus, all invariant operators of algebra $so(n+1; j)$ are polynomial and are given by (1.50).

1.3 The Cayley–Klein unitary groups and algebras

1.3.1 *Definitions, generators, commutators*

Special unitary groups $SU(n+1; j)$ are connected with complex Cayley–Klein spaces $\mathbf{C}_{n+1}(j)$ which come out from $(n+1)$-dimensional complex space $\mathbf{C}_{n+1}$ under the mapping

$$
\phi : \mathbf{C}_{n+1} \to \mathbf{C}_{n+1}(j)
$$

$$
\phi z_0^* = z_0^*, \quad \phi z_k^* = (0, k) z_k, \quad k = 1, 2, \ldots, n, \tag{1.51}
$$

where $z_0^*, z_k^* \in \mathbf{C}_{n+1}$, $z_0, z_k \in \mathbf{C}_{n+1}(j)$ are complex Cartesian coordinates; $j = (j_1, \ldots, j_n)$ where each parameter j_k takes three values: $j_k = 1, \iota_k, i$. Quadratic form $(z^*, z^*) = \sum_{m=0}^{n} |z_m^*|^2$ of the space $\mathbf{C}_{n+1}$ turns into quadratic form

$$(z, z) = |z_0|^2 + \sum_{k=1}^{n} (0, k)^2 |z_k|^2 \qquad (1.52)$$

of the space $\mathbf{C}_{n+1}(j)$ under the mapping (1.51). Here $|z_k| = (x_k^2 + y_k^2)^{1/2}$ is an absolute value (modulus) of the complex number $z_k = x_k + j y_k$, and z is complex vector: $z = (z_0, z_1, \ldots, z_n)$.

Definition of complex fiber space is similar to the real fiber space in section 1.2.3.

Definition 1.6. Group $SU(n + 1; j)$ consists of all transformations of space $\mathbf{C}_{n+1}(j)$ with unit determinant, keeping invariant the quadratic form (1.52).

In the $(k_1, k_2, \ldots, k_p)$-fiber space $\mathbf{C}_{n+1}(j)$ we have $p+1$ quadratic forms, which remain invariant under transformations of group $SU(n+1; j)$. Under transformations of group $SU(n + 1; j)$, which do not affect coordinates $z_0, z_1, \ldots, z_{k_s-1}$, the form

$$(z, z)_{s+1} = \sum_{a=k_s}^{k_{s+1}-1} (k_s, a)^2 |z_a|^2, \qquad (1.53)$$

where $s = 0, 1, \ldots, p$, $k_0 = 0$, remains invariant. For $s = p$ the summation over a goes up to n.

The mapping (1.51) induces the transition of classical group $SU(n+1)$ into group $SU(n + 1; j)$ and correspondingly, of algebra $su(n + 1)$ into algebra $su(n + 1; j)$. All $(n + 1)^2 - 1$ generators of algebra $su(n + 1)$ are Hermitian matrices. However, because the commutators for Hermitian generators are not symmetric, one usually prefers matrix generators A_{km}^*, $k, m = 0, 1, 2, \ldots, n$ of general linear algebra $gl_{n+1}(\mathbb{R})$, such that $(A_{km}^*)_{km} = 1$ and all other matrix elements vanish. (The asterisk means that A^* is a generator of a classical algebra.) The commutators of generators A^* satisfy the following relations

$$[A_{km}^*, A_{pq}^*] = \delta_{mp} A_{kq}^* - \delta_{kq} A_{pm}^*, \qquad (1.54)$$

where δ_{mp} is the Kronecker symbol. Independent Hermitian generators of algebra $su(n+1)$ are given by the equations

$$Q_{rs}^* = \frac{i}{2}(A_{rs}^* + A_{sr}^*), \quad L_{rs}^* = \frac{1}{2}(A_{sr}^* - A_{rs}^*),$$
$$P_k^* = \frac{i}{2}(A_{k-1,k-1}^* - A_{kk}^*),$$

(1.55)

where $r = 0, 1, \ldots, n-1$, $s = r+1, r+2, \ldots, n$, $k = 1, 2, \ldots, n$.

Matrix generators A^* are transformed under the mapping (1.51) as follows:

$$A_{rs}(j) = (r, s)A_{rs}^*, \quad A_{kk}(j) = A_{kk}^*.$$

(1.56)

The commutators of generators $A(j)$ can be easily found:

$$[A_{km}, A_{pq}] = (k, m)(p, q)\left(\delta_{mp}A_{kq}(k, q)^{-1} - \delta_{kp}A_{pm}(m, p)^{-1}\right).$$

(1.57)

Hermitian generators (1.55) are transformed in the same way under transition from algebra $su(n+1)$ to algebra $su(n+1; j)$. This enables us to find matrix generators of algebra $su(n+1; j)$ when group $SU(n+1; j)$ acts in the space $C_{n+1}(j)$ with named coordinates

$$Q_{rs}(j) = (r, s)Q_{rs}^*, \quad L_{rs}(j) = (r, s)L_{rs}^*, \quad P_k(j) = P_k^*.$$

(1.58)

We do not cite the commutation relations for these generators because they are cumbersome. They can be found, using (1.57).

Let us give one more realization of generators for unitary group. If group GL_{n+1} acts via left translations in the space of analytic functions on $\mathbf{C}_{n+1}$, then the generators of its algebra are $X_{ab}^* = z^{*b}\partial_a^*$, where $\partial_a^* = \frac{\partial}{\partial z^{*a}}$. Hermitian generators of algebra $su(n+1)$ can be expressed in terms of X_{ab}^* using (1.55), in which A^* must be changed for X^*. Under the mapping ϕ they are transformed according to the rule

$$Z_{ab} = (a, b)Z_{ab}^*(\phi z^*),$$

(1.59)

where $Z_{ab} = Q_{rs}, L_{rs}, P_k = P_{kk}$. Generators X_{ab}^* are transformed in a similar way, and this gives us

$$X_{kk} = z_k\partial_k, \ X_{sr} = z_r\partial_s, \ X_{rs} = (r, s)^2 z_s\partial_r,$$

(1.60)

where $k = 1, 2, \ldots, n$; $r, s = 0, 1, \ldots, n$; $r < s$.

The matrix generators (1.58) make a basis of Lie algebra $su(n+1;j)$. To the general element of the algebra

$$Z(\mathbf{u}, \mathbf{v}, \mathbf{w}; j) = \sum_{t=1}^{n(n+1)/2} (u_t Q_t(j) + v_t L_t(j)) + \sum_{k=1}^{n} w_k P_k, \qquad (1.61)$$

where index t is connected with the indices r, s, $r < s$, by relation

$$t = s + r(n-1) - \frac{r(r-1)}{2}, \qquad (1.62)$$

and the group parameters u_t, v_t, w_k are real, corresponds a finite group transformation of group $SU(n+1;j)$

$$W(\mathbf{u}, \mathbf{v}, \mathbf{w}; j) = \exp\{Z(\mathbf{u}, \mathbf{v}, \mathbf{w}; j)\}. \qquad (1.63)$$

According to Cayley–Hamilton theorem [Korn and Korn (1961)], matrix W can be algebraically expressed in terms of matrices Z^m, $m = 0, 1, 2, \ldots, n$, but one can derive it explicitly only for groups $SU(2; j_1)$ and $SU(3; j_1, j_2)$, which will be discussed in the next section.

1.3.2   *The unitary group* $SU(2; j_1)$

The group $SU(2; j_1)$ is the simplest one among unitary Cayley–Klein groups.

Definition 1.7. The set of all transformations of the space $\mathbf{C}_2(j_1)$, leaving invariant the quadratic form $|z_0|^2 + j_1^2 |z_1|^2$, make up the special unitary Cayley–Klein group $SU(2; j_1)$.

If group $SU(2; j_1)$ acts on the coordinates z_0, $j_1 z_1$ of the space $\mathbf{C}_2(j)$, then its elements are the following matrices:

$$g(j) = \begin{pmatrix} \alpha & -j\overline{\beta} \\ j\beta & \overline{\alpha} \end{pmatrix}, \quad \det g(j) = |\alpha|^2 + j^2|\beta|^2 = 1. \qquad (1.64)$$

Here the bar notes complex conjugation. Constructing generators of algebra $su(2; j_1)$ according to (1.58), we get

$$P_1 = \frac{i}{2}\begin{pmatrix} 1 & 0 \\ 0 & -1 \end{pmatrix}, \quad Q_{01} = \frac{i}{2}\begin{pmatrix} 0 & j_1 \\ j_1 & 0 \end{pmatrix}, \quad L_{01} = \frac{1}{2}\begin{pmatrix} 0 & -j_1 \\ j_1 & 0 \end{pmatrix}, \qquad (1.65)$$

and find commutation relations

$$[P_1, Q_{01}] = L_{01}, \quad [L_{01}, P_1] = Q_{01}, \quad [Q_{01}, L_{01}] = j_1^2 P_1. \qquad (1.66)$$

The generators (1.65) for $j_1 = 1$ coincide with Pauli matrices up to a factor. It is worth mentioning that under contraction $j_1 = \iota_1$, the dimension (number of linearly independent generators) of general linear group $GL(2; j_1)$ (or its algebra) diminishes as a result of the vanishing generator $A_{01}(\iota_1)$. For special unitary groups (algebras) in complex Cayley–Klein spaces, the dimension of the groups (algebras) for any (including nilpotent) values of parameters remains unchanged.

One-dimensional subgroup, corresponding to the generators (1.65), are as follows:

$$g_1(r; j_1) = \exp r Q_{01}(j_1) = \begin{pmatrix} \cos\dfrac{1}{2}j_1 r & i\sin\dfrac{1}{2}j_1 r \\ i\sin\dfrac{1}{2}j_1 r & \cos\dfrac{1}{2}j_1 r \end{pmatrix},$$

$$g_2(s; j_1) = \exp s L_{01}(j_1) = \begin{pmatrix} \cos\dfrac{1}{2}j_1 s & -\sin\dfrac{1}{2}j_1 s \\ \sin\dfrac{1}{2}j_1 s & \cos\dfrac{1}{2}j_1 s \end{pmatrix}, \qquad (1.67)$$

$$g_3(w) = \exp w P_1 = \begin{pmatrix} e^{iw/2} & 0 \\ 0 & e^{-iw/2} \end{pmatrix},$$

and with exponential mapping, we relate the matrix of finite transformation of group $SU(2; j_1)$ to the general element $Z = r Q_{01} + s L_{01} + w P_1$ of algebra $su(2; j_1)$:

$$g(\zeta, w; j_1) = \exp Z = \begin{pmatrix} \cos\dfrac{v}{2} + i\dfrac{w}{v}\sin\dfrac{v}{2} & -j_1\dfrac{\bar\zeta}{v}\sin\dfrac{v}{2} \\ j_1\dfrac{\zeta}{v}\sin\dfrac{v}{2} & \cos\dfrac{v}{2} - i\dfrac{w}{v}\sin\dfrac{v}{2} \end{pmatrix},$$

$$v^2(j_1) = w^2 + j_1^2|\zeta|^2, \quad \zeta = s + ir. \qquad (1.68)$$

In Euler parametrization [Vilenkin (1968)], transformations from group $SU(2; j_1)$ can be written as

$$g(\varphi, \theta, w; j_1) = g_3(\varphi; j_1) g_1(\theta; j_1) g_3(w; j_1)$$

$$= \begin{pmatrix} e^{i\frac{w+\varphi}{2}}\cos j_1\dfrac{\theta}{2} & e^{-i\frac{w-\varphi}{2}} i\sin j_1\dfrac{\theta}{2} \\ e^{i\frac{w-\varphi}{2}} i\sin j_1\dfrac{\theta}{2} & e^{-i\frac{w+\varphi}{2}}\cos j_1\dfrac{\theta}{2} \end{pmatrix}, \qquad (1.69)$$

where group parameters (Euler angles) are in the bounds

$$0 \leq \varphi < 2\pi, \quad -2\pi \leq w \leq 2\pi, \quad \theta \in \Theta(j) = \begin{cases} (0,\pi), & j_1 = 1 \\ (0,\infty), & j_1 = \iota \\ (-\infty,0), & j_1 = i. \end{cases} \quad (1.70)$$

Let us note that for $j_1 = 1$, matrices $g(\varphi,\theta,w;j_1)$ coincide with matrices (1.1.3–4), ch. III in [Vilenkin (1968)]; for $j_1 = i$ they coincide with the matrices (1.3.4–5), ch. VI in [Vilenkin (1968)], and for $j_1 = \iota_1$ they describe Euclidean group $SU(2;\iota_1)$ in Euler parametrization.

1.3.3 *Representations of the group $SU(2;j_1)$*

Let $\mathbf{H}(j_1) = \{f(e^{it})\}$ be a space of infinitely differentiable square-integrable functions on circle. We define representation operators corresponding to matrices (1.64) from group $SU(2;j_1)$ by the formula

$$T_\lambda(\tilde{g}(j_1))f(e^{it}) = \left(\alpha + j_1 \beta e^{-it}\right)^{\frac{\lambda}{j_1}} \left(\bar{\alpha} - j_1 \bar{\beta} e^{it}\right)^{\frac{\lambda}{j_1}} f\left(\frac{\alpha e^{it} + j_1 \beta}{\bar{\alpha} - j_1 \bar{\beta} e^{it}}\right).$$
$$(1.71)$$

For $j_1 = 1$ these operators coincide with the operators (2.1.10), ch. III in [Vilenkin (1968)], if $\lambda = l = 0, \frac{1}{2}, 1, \ldots$ and therefore describe irreducible representations of group $SU(2)$ when we restrict the space $\mathbf{H}(1)$ to the finite dimensional subspace of trigonometric polynomials $\mathbf{H}_l = \{\Phi(e^{it}) = \sum_{k=-l}^{l} a_k e^{ikt}\}$. For $j_1 = i$ the operators (1.71) coincide with the operators (2.2.5), ch. VI in [Vilenkin (1968)], if $\lambda \in C$ and describe irreducible representations of group $SU(2;i) \equiv SU(1,1)$, which are unitary for $\lambda = -\frac{i}{2} - \rho$, $\rho \in R$ (first principal series). If $j_1 = \iota_1$ in (1.71) and we take into account the restriction (1.64) on the group parameters — expressed as $|\alpha|^2 = 1$, i.e. $\alpha = e^{iw/2}$, $\beta \in C$ — then (1.71) is rewritten in the form

$$T_\lambda(g(\iota_1))f(e^{it})$$
$$= \left[1 + \iota_1 2i\mathrm{Im}(\beta)e^{-i(t+w/2)}\right]^{\frac{\lambda}{\iota_1}} f\left(e^{i(t+w)}(1 + \iota_1 2\mathrm{Re}(\beta)e^{-i(t+w/2)})\right).$$
$$(1.72)$$

Taking into account that $(1+\iota_1 a)^{\frac{\lambda}{\iota_1}} = \exp\{\frac{\lambda}{\iota_1}\ln(1+\iota_1 a)\} = \exp\{\frac{\lambda}{\iota_1}\iota_1 a)\} = \exp\{\lambda a\}$, $f(a+\iota_1 b) = f(a) + \iota_1 b f'(a)$ and omitting the nilpotent terms in

(1.72), we obtain the following expression for the representing operators of Euclidean group $SU(2; \iota_1) = M(2)$

$$T_\lambda(g(\iota_1))f(e^{it}) = e^{2i\lambda|\beta|\cos(t+p+w/2)} f\left(e^{i(t+w)}\right), \qquad (1.73)$$

where $\arg\beta = -p + \pi/2$. If we replace group parameters w, β by the new $-w, a, b$ according to the rule: $\beta = i\zeta\exp iw/2$, where $\zeta = a + ib$, $|\zeta| = r$, $\arg\zeta = \varphi$; then (1.73) takes the form

$$T_\lambda(g(\iota_1))f(e^{it}) = e^{2i\lambda r\cos(t-\varphi)} f\left(e^{i(t-w)}\right), \qquad (1.74)$$

and coincide with operators of irreducible representation of Euclidean group $M(2)$ for $2i\lambda = R$ (compare with (2.1.3) in [Vilenkin (1968)]). As far as the last ones are unitary for $R = i\rho$, $\rho \in \mathbb{R}$, then the operators (1.74) are unitary for real values of λ.

So, the formula (1.71) provides a unified description of irreducible representation operators for three groups: $SU(2)$, $SU(1,1)$ and $SU(2; \iota_1)$. The last one has the structure of the semidirect product of Abelian subgroup e^T, $T = \{Q_{01}, L_{01}\}$ and the subgroup $U_1 = \exp wP_1$, i.e. $SU(2; \iota_1) = e^T \otimes U_1$. Similar groups are called [Perroud (1983)] inhomogeneous unitary groups. Group $SU(2; \iota_1)$ is locally isomorphic to Euclidean group, which is the semidirect product of the two-dimensional translation subgroup and the plane rotation subgroup.

The unified description of the irreducible representation spaces is more complicated. For noncompact groups $SU(2; \iota_1)$ or $SU(2; i)$, the irreducible representation spaces coincide identically, $\mathbf{H}(\iota) = \mathbf{H}(i) \equiv \mathbf{H}$, converging to the space of infinitely differentiable square-integrable functions on a circle; whereas for compact group $SU(2; 1)$ the space $\mathbf{H}$ is restricted to the finite dimensional subspace $\mathbf{H}_l = \mathbf{H}(1)$ of degree l trigonometric polynomials. The reason for this is that contraction (or limit transition) is a local operation, therefore one can find the unified description of the irreducible representation operators (1.71). On the other hand, the irreducible representation space depends on global group properties (its compactness or noncompactness), therefore the structure of the irreducible representation space needs to be determined separately for each group.

The basis in the space $\mathbf{H}(j_1)$ can be written as follows $f_k^\lambda(t; j_1) = C_k^\lambda(j_1)e^{-ikt}$, where for $j_1 = \iota_1$, $i, k = 0, \pm 1, \pm 2, \ldots$, $C_k^\lambda(\iota_1) = C_k^\lambda(i) = 1$, and for $j_1 = 1$ we have $\lambda = 0, \frac{1}{2}, 1, \ldots$, $-\lambda \leq k \leq \lambda$, $C_k^\lambda(1) = [(\lambda - k)!(\lambda + k)!]^{-1/2}$. The scalar product is defined by the formula

$$\left(f_k^\lambda(t; j_1), f_m^\lambda(t; j_1) \right)$$

$$= \left(C_k^\lambda(j_1) \right)^{-2} \frac{1}{2\pi} \int_0^{2\pi} f_k^\lambda(t; j_1) \bar{f}_m^\lambda(t; j_1) dt = \delta_{km} \qquad (1.75)$$

(compare with (3.2.14), ch. III in [Vilenkin (1968)] for $j_1 = 1$, with (2.1.4), ch. IV in [Vilenkin (1968)] for $j_1 = \iota_1$ and with (2.7.4), ch. VI in [Vilenkin (1968)] for $j_1 = i$).

The operators (1.71) realize irreducible representations of groups $SU(2; j_1)$, if $\lambda = 0, \frac{1}{2}, 1, \ldots$ for $j_1 = 1$ and $\lambda \in \mathbb{C}$ for $j_1 = \iota_1, i$. These representations are unitary, if $\lambda = 0, \frac{1}{2}, 1, \ldots$ for $j_1 = 1$, $\lambda \in \mathbb{R}$; $\lambda > 0$ for $j_1 = \iota_1$ and $\lambda = -\frac{i}{2} - \rho$, $\rho \in \mathbb{R}$ for $j_1 = i$. In the last case we have irreducible representations of the first principal series for group $SU(1,1)$ (see n.7, §2, ch. VI in [Vilenkin (1968)]).

Let us consider matrix elements of the unitary irreducible representations of groups $SU(2; j_1)$ in the basis $\{f_k^\lambda(t; j_1)\}$. Using (1.71), (1.75), we obtain

$$D_{mn}^\lambda(g(j_1)) \equiv \left(T_\lambda(g(j_1)) f_n^\lambda, f_m^\lambda \right) = \frac{C_n^\lambda(j_1)}{C_m^\lambda(j_1)} \frac{1}{2\pi} \int_0^{2\pi} (\alpha + j_1 \beta e^{-it})^{\frac{\lambda}{j_1} - n}$$

$$\times (\bar{\alpha} - j_1 \bar{\beta} e^{it})^{\frac{\lambda}{j_1} + n} e^{i(m-n)t} dt. \qquad (1.76)$$

The matrix of operator $T_\lambda(g_3(w; j_1))$ is diagonal finite for $j_1 = 1$ and is an infinite matrix for $j_1 = \iota_1, i$, the main diagonal of which is composed of numbers $\exp(-imw)$. Therefore, from Euler decomposition (1.69) we have

$$D_{mn}^\lambda(g(\varphi, \theta, \omega; j_1)) = e^{-i(m\varphi + n\omega)} P_{mn}^\lambda(\theta; j_1), \qquad (1.77)$$

where $P_{mn}^\lambda(\theta; j_1)$ is the matrix element corresponding to subgroup $g_1(\theta; j_1)$:

$$P_{mn}^\lambda(\theta; j_1) = \frac{C_n^\lambda(j_1)}{C_m^\lambda(j_1)} \frac{1}{2\pi} \int_0^{2\pi} \left(\cos j_1 \frac{\theta}{2} + e^{-it} i \sin j_1 \frac{\theta}{2} \right)^{\frac{\lambda}{j_1} - n}$$

$$\times \left(\cos j_1 \frac{\theta}{2} + e^{it} i \sin j_1 \frac{\theta}{2} \right)^{\frac{\lambda}{j_1} + n} e^{i(m-n)t} dt. \qquad (1.78)$$

For $j_1 = 1$ the expression (1.78) coincides with determining function $P_{mn}^l(\cos \theta)$ (3.4.4), ch. III in [Vilenkin (1968)], which is the particular case of Jacobi polynomials $P_k^{(a,b)}(z)$ with integer a and b. For $j_1 = i$ the expression (1.78) coincides with (3.2.6), ch. VI in [Vilenkin (1968)], determining Jacobi function $\mathbf{P}^l_{mn}(\cosh\theta)$. At last, for $j_1 = \iota_1$ we have $C^\lambda_n(\iota_1) = C^\lambda_m(\iota_1) = 1$ and (1.78) can be rewritten in the form

$$P^\lambda_{mn}(\theta;\iota_1) = \frac{1}{2\pi}\int_0^{2\pi}(1+\iota_1 i\theta\cos t)^{\frac{\lambda}{\iota_1}}\,e^{i(m-n)t}dt$$

$$= \frac{1}{2\pi}\int_0^{2\pi}e^{i\lambda\theta\cos t+i(m-n)t}dt = i^{n-m}J_{n-m}(\lambda\theta), \quad (1.79)$$

coinciding with the matrix element of Euclidean group (see §3, ch. IV in [Vilenkin (1968)]). Here $J_n(x)$ is the Bessel function of n-th order. So, the formula (1.78) gives the unified expression of matrix elements for groups $SU(2), M(2)$ and $SU(1,1)$.

The group law of composition generates functional relations for matrix elements of representation. These relations can be written in a unified form as well. Let us consider particularly the summation theorem for functions $P^\lambda_{mn}(\theta;\iota_1)$. From the relation $g(\varphi,\theta,\omega) = g_1(0,\theta_1,0)g_2(\varphi_2,\theta_2,0)$ we find Euler parameters φ,θ,ω as functions of $\theta_1,\theta_2,\varphi_2$ in the form

$$\cos j_1\theta = \cos j_1\theta_1\cos j_1\theta_2 - \sin j_1\theta_1\sin j_1\theta_2\cos\varphi_2,$$

$$e^{i\varphi} = \frac{1}{\sin j_1\theta}(\sin j_1\theta_1\cos j_1\theta_2 + \cos j_1\theta_1\sin j_1\theta_2\cos\varphi$$

$$+ i\sin j_1\theta_2\sin\varphi_2),$$

$$e^{i(\varphi+\omega)/2} = \frac{1}{\cos j_1\frac{\theta}{2}}\left[\cos j_1\frac{\theta_1}{2}\cos j_1\frac{\theta_2}{2}e^{i\varphi_2/2}\right.$$

$$\left. - \sin j_1\frac{\theta_1}{2}\sin j_1\frac{\theta_2}{2}e^{-i\varphi_2/2}\right], \quad (1.80)$$

$$\frac{1}{j_1^2}\sin^2 j_1\theta = \frac{1}{j_1^2}\sin^2 j_1\theta_1 + \frac{1}{j_1^2}\sin^2 j_1\theta_2$$

$$- \frac{1}{j_1^2}\sin^2 j_1\theta_1\sin^2 j_1\theta_2(1+\cos^2\varphi_2)$$

$$+ \frac{1}{2j_1^2}\sin 2j_1\theta_1\sin 2j_1\theta_2\cos\varphi_2.$$

As far as $j_1 = \iota_1$, cosine is equal to one: $\cos\iota_1\theta = 1$. Then the first equation in (1.80) degenerates into the identity $1 \equiv 1$, therefore we rewrite it in the equivalent form through sines (the last equation in (1.80)). For $j_1 = 1$ the equations (1.80) coincide with the expressions (4.1.6), ch. III in [Vilenkin (1968)]; for $j_1 = i$ they coincide with the expressions (4.1.4), ch. VI in [Vilenkin (1968)]; and for $j_1 = \iota_1$ they take the form

$$\theta^2 = \theta_1^2 \theta_2^2 + 2\theta_1\theta_2 \cos\varphi_2, \quad e^{i\varphi} = \frac{1}{\theta}\left(\theta_1 + \theta_2 e^{i\varphi_2}\right),$$

$$e^{i(\varphi+\omega)/2} = e^{i\varphi_2/2} \tag{1.81}$$

which agree with expressions (4.1.2), ch. IV in [Vilenkin (1968)]. From the equation $T_\lambda(g) = T_\lambda(g_1)T_\lambda(g_2)$, for operators of representation, it follows for matrix elements that:

$$D_{mn}^\lambda(g(\varphi,\theta,\omega)) = \sum_k D_{mk}^\lambda(g_1(0,\theta_1,0))D_{kn}^\lambda(g_2(\varphi_2,\theta_2,0)), \tag{1.82}$$

where the summation of k is taken from $-\lambda$ to λ for $j_1 = 1$ and from $-\infty$ to ∞ for $j_1 = \iota_1, i$. Taking into account of (1.77)

$$D_{mk}^\lambda(g_1) = P_{mk}^\lambda(\theta_1; j_1), \quad D_{kn}^\lambda(g_2) = e^{-ik\varphi_2} P_{kn}^\lambda(\theta_2; j_1),$$

$$D_{mn}^\lambda(g) = e^{-i(m\varphi+n\omega)} P_{mn}^\lambda(\theta; j_1), \tag{1.83}$$

we obtain the summation theorem for functions $P_{mn}^\lambda(\theta; j_1)$ in the form

$$e^{-i(m\varphi+n\omega)} P_{mn}^\lambda(\theta; j_1) = \sum_k e^{-ik\varphi_2} P_{mk}^\lambda(\theta_1; j_1) P_{kn}^\lambda(\theta_2; j_1). \tag{1.84}$$

This expression agrees identically with the summation theorem for Jacobi polynomials (4.1.7), ch. III in [Vilenkin (1968)] and Jacobi functions (4.1.6), ch. VI in [Vilenkin (1968)] for $j_1 = 1$ and $j_1 = i$ respectively. For $j_1 = \iota_1$, the expression (1.84) in consideration of (1.79), takes the form

$$e^{in\varphi} J_n(\lambda\theta) = \sum_{k=-\infty}^{\infty} e^{ik\varphi_2} J_{n-k}(\lambda\theta_1) J_k(\lambda\theta_2), \tag{1.85}$$

where parameters θ, φ are connected with parameters $\theta_1, \theta_2, \varphi_2$ by (1.81), and gives the summation theorem for Bessel functions (4.1.3), ch. IV in [Vilenkin (1968)].

In much the same way, we can write down uniformly the other properties of the matrix elements such as their generating function $P_{mn}^\lambda(\theta; j_1)$. Following from the relation (1.78) and Fourier transformation, we arrive at the equation

$$F(\theta, e^{-it}; j_1) \equiv C_n^\lambda(j_1) \left(\cos j_1 \frac{\theta}{2} + e^{-it} i \sin j_1 \frac{\theta}{2} \right)^{\frac{\lambda}{j_1} - n}$$

$$\times \left(\cos j_1 \frac{\theta}{2} + e^{it} i \sin j_1 \frac{\theta}{2} \right)^{\frac{\lambda}{j_1} + n} e^{-int}$$

$$= \sum_m P_{mn}^\lambda(\theta; j_1) C_m^\lambda(j_1) e^{-imt}, \tag{1.86}$$

where the summation over m is taken from $-\lambda$ to λ for $j_1 = 1$ and from $-\infty$ to ∞ for $j_1 = \iota_1, i$. The coefficients $C_m^\lambda(j_1)$ have been defined earlier. The formula (1.86) shows that $F(\theta, e^{-it}; j_1)$ is a generating function of $P_{mn}^\lambda(\theta; j_1)$. Introducing a new variable $w = e^{it}$ for $j_1 = 1$, we obtain the expression from (1.86), which coincides with the generating function of Jacobi polynomials $P_{mn}^l(\cos \theta)$ (see (5.1.3), ch. III in [Vilenkin (1968)]). For $j_1 = i$, and denoting $z = e^{-it}$, we obtain the generating function of Jacobi functions $\mathbf{P}_{mn}^\lambda(\cosh \theta)$ from (1.86) (compare with (4.6.2), ch. VI in [Vilenkin (1968)]). For $j_1 = \iota_1$, (1.86) is the right generating function for Bessel functions (see (4.5.2), ch. IV in [Vilenkin (1968)]).

Thus, the unified description of Cayley–Klein groups $SU(2; j_1)$ generates the unified description of their representing operators (1.71) and the matrix elements (1.73) of these operators. The matrix elements (1.75) of the operators, corresponding to subgroup $g_1(\theta; j_1)$, for $j_1 = 1$ give Jacobi polynomials $P_{mn}^l(\cos \theta)$, the analytic continuation of which (for $j_1 = i$), i.e. $l \to \lambda \in C$, $\theta \to i\theta$, results in Jacobi functions $\mathbf{P}_{mn}^\lambda(\cosh \theta)$. The contraction of $SU(2)$ to Euclidean group $SU(2; \iota_1) = M(2)$ induces the limit transitions between the matrix elements (special functions of mathematical physics), namely: Jacobi polynomials $P_{mn}^l(\cos \theta)$ turn into Bessel functions $J_{n-m}(\lambda \theta)$ (compare [Vilenkin (1968)], ch. IV, §7, n.2, where the limit transition $P_{mn}^l \left(\cos \frac{r}{l} \right) \to J_{n-m}(r)$ for $l \to \infty$ was considered). The properties of special functions, which have group nature, such as the summation theorem (1.85), the generating function (1.86) and, evidently, other properties admit the unified description as well.

But properties, which are determined by the global structure of groups, need to be investigated independently for each individual group. These are choice of spaces, where representations are irreducible, and the choice of values in parameters λ, marking representations, when the last ones are unitary.

1.3.4 *The unitary group $SU(3;j)$*

Group $SU(3;j)$, $j = (j_1, j_2)$ consists of transformations, preserving the quadratic form $(\mathbf{z}, \mathbf{z}) = |z_0|^2 + j_1^2|z_1|^2 + j_1^2 j_2^2 |z_2|^2$. The matrix generators of general linear algebra $gl_3(j)$, according to (1.56), are as follows

$$A_{00} = \begin{pmatrix} 1 & 0 & 0 \\ 0 & 0 & 0 \\ 0 & 0 & 0 \end{pmatrix}, \quad A_{11} = \begin{pmatrix} 0 & 0 & 0 \\ 0 & 1 & 0 \\ 0 & 0 & 0 \end{pmatrix}, \quad A_{22} = \begin{pmatrix} 0 & 0 & 0 \\ 0 & 0 & 0 \\ 0 & 0 & 1 \end{pmatrix},$$

$$A_{10} = \begin{pmatrix} 0 & 0 & 0 \\ j_1 & 0 & 0 \\ 0 & 0 & 0 \end{pmatrix}, \quad A_{01} = \begin{pmatrix} 0 & j_1 & 0 \\ 0 & 0 & 0 \\ 0 & 0 & 0 \end{pmatrix}, \quad A_{20} = \begin{pmatrix} 0 & 0 & 0 \\ 0 & 0 & 0 \\ j_1 j_2 & 0 & 0 \end{pmatrix},$$

$$A_{02} = \begin{pmatrix} 0 & 0 & j_1 j_2 \\ 0 & 0 & 0 \\ 0 & 0 & 0 \end{pmatrix}, \quad A_{21} = \begin{pmatrix} 0 & 0 & 0 \\ 0 & 0 & 0 \\ 0 & j_2 & 0 \end{pmatrix}, \quad A_{12} = \begin{pmatrix} 0 & 0 & 0 \\ 0 & 0 & j_2 \\ 0 & 0 & 0 \end{pmatrix}.$$

$$(1.87)$$

Let us write down the generators of algebra $su(3;j)$ constructed from the generators (1.87) according to the formulae (1.58)

$$P_1 = \frac{i}{2} \begin{pmatrix} 1 & 0 & 0 \\ 0 & -1 & 0 \\ 0 & 0 & 0 \end{pmatrix}, \quad P_2 = \frac{i}{2} \begin{pmatrix} 0 & 0 & 0 \\ 0 & 1 & 0 \\ 0 & 0 & -1 \end{pmatrix},$$

$$Q_1 = \frac{i}{2} \begin{pmatrix} 0 & j_1 & 0 \\ j_1 & 0 & 0 \\ 0 & 0 & 0 \end{pmatrix}, \quad L_1 = \frac{1}{2} \begin{pmatrix} 0 & -j_1 & 0 \\ j_1 & 0 & 0 \\ 0 & 0 & 0 \end{pmatrix},$$

$$Q_2 = \frac{i}{2} \begin{pmatrix} 0 & 0 & j_1 j_2 \\ 0 & 0 & 0 \\ j_1 j_2 & 0 & 0 \end{pmatrix}, \quad L_2 = \frac{1}{2} \begin{pmatrix} 0 & 0 & -j_1 j_2 \\ 0 & 0 & 0 \\ j_1 j_2 & 0 & 0 \end{pmatrix},$$

$$Q_3 = \frac{i}{2} \begin{pmatrix} 0 & 0 & 0 \\ 0 & 0 & j_2 \\ 0 & j_2 & 0 \end{pmatrix}, \quad L_3 = \frac{1}{2} \begin{pmatrix} 0 & 0 & 0 \\ 0 & 0 & -j_2 \\ 0 & j_2 & 0 \end{pmatrix},$$

$$(1.88)$$

where $Q_k \equiv Q_{0k}$, $L_k \equiv L_{0k}$, $k = 1, 2$, $Q_3 \equiv Q_{12}$ $L_3 \equiv L_{12}$. These generators satisfy the following commutation relations:

$$[P_1, P_2] = 0, \quad [P_1, Q_1] = L_1, \quad [P_1, L_1] = -Q_1, \quad [P_1, Q_2] = \frac{1}{2} L_2,$$

$$[P_1, L_2] = -\frac{1}{2} Q_2, \quad [P_1, Q_3] = -\frac{1}{2} L_3, \quad [P_1, L_3] = \frac{1}{2} Q_3,$$

$$[P_2, Q_1] = -\frac{1}{2} L_1, \quad [P_2, L_1] = \frac{1}{2} Q_1, \quad [P_2, Q_2] = \frac{1}{2} L_2,$$

$$[P_2, L_2] = -\frac{1}{2}Q_2, \quad [P_2, Q_3] = L_3, \quad [P_2, L_3] = -Q_3,$$

$$[Q_1, L_1] = j_1^2, \quad [Q_2, L_2] = j_1^2 j_2^2 (P_1 + P_2), \quad [Q_3, L_3] = j_2^2 P_2,$$

$$[Q_1, L_2] = -\frac{j_1^2}{2}Q_3, \quad [Q_2, L_3] = \frac{j_2^2}{2}Q_1, \quad [L_1, Q_2] = \frac{j_1^2}{2}Q_3,$$

$$[L_2, Q_3] = -\frac{j_2^2}{2}Q_1, \quad [Q_1, L_3] = -\frac{1}{2}Q_2, \quad [Q_3, L_1] = \frac{1}{2}Q_2,$$

$$[Q_1, Q_2] = \frac{j_1^2}{2}L_3, \quad [Q_1, Q_3] = \frac{1}{2}L_2, \quad [Q_2, Q_3] = \frac{j_2^2}{2}L_1,$$

$$[L_1, L_2] = \frac{j_1^2}{2}L_3, \quad [L_1, L_3] = -\frac{1}{2}L_2, \quad [L_2, L_3] = \frac{j_2^2}{2}L_1. \tag{1.89}$$

Contractions change the structure of groups (algebras). Let us put $j_1 = \iota_1$ in (1.89). The obtained commutation relations show that simple classical algebra $su(3)$ acquires the structure of the semidirect sum: $su(3; \iota_1, j_2) = T \oplus u(2; j_2)$, where the commutative ideal T is spanned over the generators Q_1, L_1, Q_2, L_2 and subalgebra $u(2; j_2)$ spanned over the generators P_1, P_2, Q_3, L_3 is the Lie algebra of unitary group in complex Cayley–Klein space. It follows from (1.89) for $j_1 = \iota_1$ that $[T, u(2; j_2)] \subset T$, as it can be expected for the semidirect sum of algebras. Group $SU(3; \iota_1, j_2)$ has a structure of the semidirect product $SU(3; \iota_1, j_2) = \exp(T) \otimes U(2; j_2)$ and is a so-called inhomogeneous unitary group [Perroud (1983)].

In relation to a general element of algebra $su(3; j)$

$$Z(\mathbf{u}, \mathbf{v}, \mathbf{w}; j) = \sum_{k=1}^{3} (u_k Q_k + v_k L_k) + w_1 P_1 + w_2 P_2$$

$$= \frac{1}{2}\begin{pmatrix} iw_1 & -j_1(v_1 - iu_1) & -j_1 j_2(v_2 - iu_2) \\ j_1(v_1 + iu_1) & i(w_2 - w_1) & -j_2(v_3 - iu_3) \\ j_1 j_2(v_2 + iu_2) & j_2(v_3 + iu_3) & -iw_2 \end{pmatrix}$$

$$= \frac{1}{2}\begin{pmatrix} iw_1 & -j_1 \bar{t}_1 & -j_1 j_2 \bar{t}_2 \\ j_1 t_1 & i(w_2 - w_1) & -j_2 \bar{t}_3 \\ j_1 j_2 t_2 & j_2 t_3 & -iw_2 \end{pmatrix}, \tag{1.90}$$

where $t_k = v_k + iu_k$, $k = 1, 2, 3$ are complex parameters, and the bar means the complex conjugation, the finite group transformation of $SU(3; j)$ acting on vectors with the named components is:

$$W(\mathbf{t}, \mathbf{w}; j) = \exp\{Z(\mathbf{t}, \mathbf{w}; j)\} \in SU(3; j). \tag{1.91}$$

To find the matrix W, let us appeal to Cayley–Hamilton theorem. The characteristic equation $\det(Z - hE_3) = 0$ for the matrix Z is of a cubic nature:

$$h^3 + ph + q = 0,$$
$$p = w_1^2 - w_1 w_2 + w_2^2 + |t|^2(j),$$
$$|t|^2(j) = j_1^2|t_1|^2 + j_1^2 j_2^2|t_2|^2 + j_2^2|t_3|^2, \tag{1.92}$$
$$q = -iw_1 w_2(w_2 - w_1) + iw_2 j_1^2|t_1|^2 - i(w_2 - w_1)j_1^2 j_2^2|t_2|^2$$
$$- iw_1 j_2^2|t_3|^2 + 2ij_1^2 j_2^2 \mathrm{Im}(t_1 \bar{t}_2 t_3).$$

Its roots are

$$h_k = \sqrt[3]{-\frac{q}{2} + \sqrt{\left(\frac{q}{2}\right)^2 + \left(\frac{p}{3}\right)^3}}$$
$$+ \sqrt[3]{-\frac{q}{2} - \sqrt{\left(\frac{q}{2}\right)^2 + \left(\frac{p}{3}\right)^3}} = h_k' + h_k'', \tag{1.93}$$

where $h_k' h_k'' = -p/3$, and index $k = 1, 2, 3$ enumerates three different cubic roots in the first summand. Using Cayley–Hamilton theorem, we find

$$W(\mathbf{t}, \mathbf{w}; j) = AE_3 - BZ + CZ^2, \tag{1.94}$$

$$4Z^2 = \begin{pmatrix} -w_1^2 - j_1^2|t_1|^2 - j_1^2 j_2^2|t_2|^2 & -j_1(iw_2\bar{t}_1 + j_2^2\bar{t}_2 t_3) \\ j_1(iw_2 t_1 - j_2^2 t_2\bar{t}_3) & -(w_2 - w_1)^2 - j_1^2|t_1|^2 - j_2^2|t_3|^2 \\ j_1 j_2(-i(w_2 - w_1)t_2 + t_1 t_3) & -j_2(iw_1 t_3 + j_1^2\bar{t}_1 t_2) \end{pmatrix}$$

$$\begin{pmatrix} j_1 j_2(i(w_2 - w_1)\bar{t}_2 + \bar{t}_1 \bar{t}_3) \\ j_2(iw_1\bar{t}_3 - j_1^2 t_1\bar{t}_2) \\ -w_2^2 - j_1^2 j_2^2|t_2|^2 - j_2^2|t_3|^2 \end{pmatrix}, \tag{1.95}$$

and the functions A, B, C can be expressed in terms of roots h_k of the characteristic equation via the relations

$$A = \{h_2 h_3(h_2 - h_3)\exp(h_1) - h_1 h_3(h_1 - h_3)\exp(h_2)$$
$$+ h_1 h_2(h_1 - h_2)\exp(h_3)\}/D,$$
$$B = \{(h_2^2 - h_3^2)\exp(h_1) - (h_1^2 - h_3^2)\exp(h_2)$$
$$+ (h_1^2 - h_2^2)\exp(h_3)\}/D, \tag{1.96}$$
$$C = \{(h_2 - h_3)\exp(h_1) - (h_1 - h_3)\exp(h_2)$$
$$+ (h_1 - h_2)\exp(h_3)\}/D,$$
$$D = (h_1 - h_2)(h_1 - h_3)(h_2 - h_3).$$

As $\operatorname{Tr} Z = 0$, $\det W = 1$ and we have the finite group transformation written in terms of coordinates of the second kind.

1.3.5 Invariant operators

Algebra $su(n+1)$ has n independent Casimir operators

$$C_p^* = \sum_{k_0,\ldots,k_p=0}^{n} A_{k_o k_1}^* A_{k_1 k_2}^* \cdots A_{k_p k_o}^*, \quad p = 1, 2, \ldots, n. \qquad (1.97)$$

To find the invariant operators of algebra $su(n+1;j)$, we shall use the method described in section 1.2.2. To derive the rule of transformation for Casimir operators under the transition from algebra $su(n+1)$ to algebra $su(n+1;j)$, let us substitute the generators A^* in (1.97) for their expressions in terms of generators $A(j)$ according to (1.56) and multiply the obtained (singulary transformed according to the terminology of Inönü and Wigner (1953)) operator, which we shall denote as $C_p^*(\rightarrow)$, by the inverse of the most singular coefficient for the terms in (1.97). For even $p = 2q$ the most singular coefficient, equal to $(0, n)^{-2q}$, is that of the summand $A_{0n} A_{n0} \cdots A_{n0} A_{0n}$, and for odd $p = 2q - 1$ the same coefficient is that of the summand $A_{0n} A_{n0} \cdots A_{0n} A_{n0}$. For this reason in accordance with (1.34), Casimir operators are transformed as follows

$$C_p(j) = (0, n)^{2q} C_p^*(\rightarrow), \qquad (1.98)$$

where $p = 2q$ or $p = 2q - 1$. Using (1.97) and (1.98), we find the invariant operators of algebra $su(n+1;j)$

$$C_p(j) = \sum_{k_o,\ldots,k_p=0}^{n} A_{k_0 k_1} A_{k_1 k_2} \cdots A_{k_p k_0} \frac{(0, n)^{2q}}{(k_0, k_p)} \prod_{m=0}^{p-1} (k_m, k_{m+1})^{-1}, \qquad (1.99)$$

where $p = 1, 2, \ldots, n$.

In [Gromov (1984a)] the following theorem is proved using the same methods as in the case of orthogonal algebras.

Theorem 1.2. *For any set of values of parameters j the number of invariant operators of algebra $su(n+1;j)$ does not exceed n.*

We shall omit the proof because it is cumbersome. The operators (1.99) are independent for $p = 1, 2, \ldots, n$; for this reason, algebra $su(n+1;j)$ has just n invariant operators, explicitly given by (1.99).

1.4 Classification of transitions between the Cayley–Klein spaces and groups

In the previous sections we have found orthogonal and unitary groups in Cayley–Klein spaces and shown that their generators, Casimir operators and other algebraic constructions can be obtained by transformation of the corresponding constructions for classical groups. Such approach is natural and is justified by the fact that classical groups and their characteristic algebraical constructions are well studied. But is such approach the only one? Is it possible to take one of the groups in Cayley–Klein space as the initial one? The positive answer to the last question is given by the following theorem [Gromov (1990a)] on the structure of transitions between groups.

Let us define (formally) the transition from the space $\mathbf{C}_{n+1}(j)$ and the generators $Z_{ab}(\mathbf{z}; j)$ of unitary group $SU(n + 1; j)$ to the space $\mathbf{C}_{n+1}(j')$ and the generators $Z_{ab}(\mathbf{z}'; j')$ via transformations obtained from (1.51) and (1.59), with the substitution of parameters j_k for $j'_k j_k^{-1}$ in the latter:

$$\phi' : \mathbf{C}_{n+1}(j) \to \mathbf{C}_{n+1}(j')$$

$$\phi' z_0^* = z_0', \quad \phi' z_k^* = z_k' \prod_{m=1}^{k} j'_m j_m^{-1}, \quad k = 1, 2, \ldots, n, \qquad (1.100)$$

$$Z_{ab}(\mathbf{z}'; j') = \left(\prod_{l=1+\min(a,b)}^{\max(a,b)} j'_l j_l^{-1} \right) Z_{ab}(\phi' \mathbf{z}; j).$$

The inverse transitions can be obtained from (1.100) by the change of the dashed parameters j' for the undashed parameters j and vice versa. Applying (1.100) to the quadratic form (1.52) and the generators (1.60), we obtain

$$(\mathbf{z}', \mathbf{z}) = |z_0'|^2 + \sum_{k=1}^{n} |z_k'|^2 \prod_{m=1}^{k} j_m'^2,$$

$$X_{kk} = z_k' \partial_k', \quad X_{sr} = z_r' \partial_s', \quad X_{rs} = \left(\prod_{l=1+r}^{s} j_l'^2 \right) z_s' \partial_r', \qquad (1.101)$$

i.e. quadratic form in space $\mathbf{C}_{n+1}(j')$ and generators of group $SU(n+1; j')$.

However, the constructed transitions do not make sense for all groups and spaces, because for the nilpotent values of parameters j the expressions ι_k^{-1} and $\iota_m \cdot \iota_k^{-1}$ for $k \neq m$ are not defined. We have defined in section 1.1 only the expressions $\iota_k \cdot \iota_k^{-1} = 1$, $k = 1, 2, \ldots, n$. So if $j_k = \iota_k$ for some k, then the transformations (1.100) will be defined and give us (1.101) provided that the dashed parameter with the same number is equal to the same nilpotent number, i.e. $j'_k = \iota_k$.

The transitions from space $\mathbf{R}_{n+1}(j)$ to space $\mathbf{R}_{n+1}(j')$, and from groups $SO(n+1;j)$, $Sp(n;j)$ to groups $SO(n+1;j')$, $Sp(n;j')$ can be correspondingly obtained from the transition (1.44) and (1.30) respectively by the same substitution of parameters j_k for $j'_k j_k^{-1}$. (Symplectic Cayley–Klein groups will be considered in chapter 3. Similarly, these relations are permitted with justification. Let us introduce the notations: $G(j) = SO(n+1;j)$, $SU(n+1;j)$, $Sp(n;j)$, $\mathbf{R}(j) = \mathbf{R}_{n+1}(j)$, $\mathbf{C}_{n+1}(j)$, $\mathbf{R}_n(j) \times \mathbf{R}_n(j)$ and denote the transformation of group generators by the symbol $\Phi G(j) = G(j')$. Easy analysis of the transformations (1.100) and their inverse — from the point of view of the admissibility of transitions [Gromov (1990a)] — implies the following theorem.

Theorem 1.3. On classification of transitions. *I. Let $G(j)$ be a group in non-fiber space $\mathbf{R}(j)$ and $G(j')$ be a group in arbitrary space $\mathbf{R}(j')$; then, $G(j') = \Psi G(j)$. If $\mathbf{R}(j')$ is a non-fiber space, Ψ is a one-to-one mapping and $G(j) = \Psi^{-1} G(j')$.*

II. Let $G(j)$ be a group in $(k_1, k_2, \ldots, k_p)$-fiber space $\mathbf{R}(j)$ and $G(j')$ be a group in $(m_1, m_2, \ldots, m_q)$-fiber space $\mathbf{R}(j')$; then, $G(j') = \Psi G(j)$ if the set of integers $(k_1, \ldots, k_p)$ is involved in the set of numbers $(m_1, \ldots, m_q)$. The inverse transition $G(j) = \Phi^{-1} G(j')$ is valid if and only if $p = q$, $k_1 = m_1, \ldots, k_p = m_p$.

It follows from the theorem that the group $G(j)$ for any set of values of the parameters j can be obtained not only from a classical group, but from a group in an arbitrary non-fiber Cayley–Klein space, i.e. from pseudoorthogonal, pseudounitary or pseudosymplectic groups. Naturally, the transitions between other algebraic constructions, in particular between Casimir operators, are described by this theorem as well.

Chapter 2

Space–time models

The kinematics groups and algebras are regarded in this chapter. The interpretations of kinematics as spaces of constant curvature are given. The non-relativistic and exotic Carroll kinematics are studied in detail.

2.1 Kinematics groups

Possible kinematic groups, i.e. groups of motion for four-dimensional models of space-time (kinematics), satisfying natural physical postulates — (1) space is isotropic; (2) spatial property of being even inversion of time are automorphisms of kinematic groups and (3) boosts (rotations in spatial-temporal plane) make noncompact subgroups — are described by Bacry and Levy-Leblond (1968). The paper [Bacry and Nuyts (1986)] rejected postulates (2) and (3) and obtained a wider set of groups with spatial isotropy. Now we shall bring the geometric interpretation of kinematics [Gromov (1990)].

All kinematic groups are ten-dimensional; for this reason, kinematics from the geometrical point of view should be among four-dimensional maximally homogeneous spaces — spaces of constant curvature, which groups of motions are of dimension 10. These spaces are realized on the connected components of the sphere

$$\mathbf{S}_4(j) = \left\{ x_0^2 + \sum_{k=1}^{4} (0,k)^2 x_k^2 = 1 \right\}. \tag{2.1}$$

Let us introduce internal (Beltramian) coordinates $\xi_k = x_k/x_0$, $k = 1,2,3,4$ on $\mathbf{S}_4(j)$. The generators (1.31) of group $SO(4;j)$ can be expressed in terms of the internal coordinates ξ via formulas

$$X_{0s}(\xi) = -\partial_1 - (0,s)^2 \xi_s \sum_{k=1}^{4} \xi_k \partial_k, \quad \partial_k = \partial/\partial\xi_k,$$

$$X_{rs}(u) = -\xi_r \partial_s + (r,s)^2 \xi_s \partial_r, \quad r < s, \ r,s = 1,2,3,4 \quad (2.2)$$

and satisfy the commutation relations (1.45). The generator $X_{0s}(u)$ has a meaning of generator for translation along the s-th Beltrami axis, and $X_{rs}(u)$ is the generator of rotation in two-dimensional plane $\{\xi_r, \xi_s\}$.

Physical postulates (1)–(3) can be expressed in terms of parameters j. Postulate (1) means that under the transformations (1.44) three Beltrami coordinates should be multiplied by the same quantity and interpreted as a temporal axis of kinematics. It is possible in two cases:

(A) for $j_3 = j_4 = 1$, when coordinates ξ_2, ξ_3, ξ_4 are multiplied by the product $j_1 j_2$ and called spatial, and ξ_1 is multiplied by j_1 and called temporal;

(B) for $j_2 = j_3 = 1$, when the spatial coordinates $\xi_k = r_k$, $k = 1,2,3$ are multiplied by j_1, and temporal coordinate $\xi_4 = t$ is multiplied by the product $j_1 j_4$.

Postulate (3) imposes restrictions on the character of rotations in two-dimensional planes, spanned over temporal and spatial axes of kinematics, requiring these rotations to be Lorentzian and Galilean. In terms of parameters j, this gives $j_2 = \iota_2, i$ in the case (A) and $j_4 = \iota_4, i$ in the case (B). The requirements of postulate (2) can be taken into account by the definition of space with the constant curvature as a connected component of the sphere (2.1).

In the case (A) the kinematic generators H, $\mathbf{P} = (P_1, P_2, P_3)$ (spatial-temporal translations), $\mathbf{J} = (J_1, J_2, J_3)$ (rotations), $\mathbf{K} = (K_1, K_2, K_3)$ (boosts) are expressed in terms of generators (2.2) in accordance with the aforementioned interpretation by the relations $H = -X_{01}$, $P_k = -X_{0,k+1}$, $K_k = -X_{1,k+1}$, $J_1 = X_{34}$, $J_2 = -X_{24}$, $J_3 = X_{23}$, $k = 1,2,3$, and satisfy the commutation relations

$$[H, \mathbf{J}] = 0, \quad [H, \mathbf{K}] = \mathbf{P}, \quad [H, \mathbf{P}] = -j_1^2 \mathbf{K}$$

$$[\mathbf{P}, \mathbf{P}] = j_1^2 j_2^2 \mathbf{J}, \quad [\mathbf{K}, \mathbf{K}] = j_2^2 \mathbf{J}, \quad [P_k, K_l] = -j_2^2 \delta_{kl} H. \quad (2.3)$$

Here $[\mathbf{X}, \mathbf{Y}] = \mathbf{Z}$ means $[X_k, Y_l] = e_{klm} Z_m$, where e_{klm} is the antisymmetric unit tensor. The spaces of constant curvature $\mathbf{S}_4(j_1, j_2, 1, 1) \equiv \mathbf{S}_4(j_1, j_2)$,

$j_1 = 1, \iota_1, i$, $j_2 = \iota_2, i$ are shown in Fig. 1.2 (see section 1.2.2), where the spatial axis r should be imagined as a three-dimensional space. Semispherical group $SO(5; 1, \iota_2)$ and semihyperbolic group $SO(5; i, \iota_2)$ correspond to Newton groups $N_\pm$ (sometimes the latter are called Hooke groups). The interpretation of other groups is well known.

In the case of (B) the temporal and spatial axes of kinematics are expressed in another way in terms of Beltramian coordinates of space with the constant curvature; correspondingly, the geometrical generators $X(\xi)$ obtain another kinematic interpretation: $H = X_{04}$, $P_k = -X_{0k}$, $K_k = X_{k4}$, $J_1 = X_{23}$, $J_2 = -X_{13}$, $J_3 = X_{12}$ and satisfy the commutation relations

$$[\mathbf{J}, \mathbf{J}] = \mathbf{J}, \quad [\mathbf{J}, \mathbf{P}] = \mathbf{P}, \quad [\mathbf{J}, \mathbf{K}] = \mathbf{K},$$

$$[H, \mathbf{J}] = 0, \quad [H, \mathbf{K}] = -j_4^2 \mathbf{P}, \quad [H, \mathbf{P}] = j_1^2 \mathbf{K},$$

$$[\mathbf{P}, \mathbf{P}] = j_1^2 \mathbf{J}, \quad [\mathbf{K}, \mathbf{K}] = j_4^2 \mathbf{J}, \quad [P_k, K_l] = \delta_{kl} H. \tag{2.4}$$

The value of parameter $j_4 = i$, as it can be readily understood, does not lead to new kinematics because $SO(5; j_1, 1, 1, i)$ is the de Sitter group, Poincaré group and anti-de Sitter group for $j_1 = 1, \iota_1$ and i respectively.

Kinematic Carroll group of motions of the flat Carroll space, first described in physical terms by J.-M. Levy-Leblond (1965), corresponds to the values of parameters $j_1 = \iota_1$, $j_4 = \iota_4$. Comparing the commutators (2.4) with the commutators in the paper [Bacry and Levy-Leblond (1968)], we find that group $SO(5; 1, 1, 1, \iota_4)$ coincides with kinematic group $ISO(4)$, and group $SO(5; i, 1, 1, \iota_4)$ is "para-Poincaré" group P'. As parameter j_1 determines the sign of the space curvature (curvature is positive for $j_1 = 1$, zero for $j_1 = \iota_1$ and negative for $j_1 = i$) we conclude that group $SO(5; 1, 1, 1, \iota_4)$ (or $ISO(4)$) is the group of motions of Carroll kinematics with a positive curvature while group $SO(5; 1, 1, 1, \iota_4)$ (or P') is the group of motions of Carroll kinematics with a negative curvature. Such an interpretation of kinematic groups $ISO(4)$ and P', as far as it can be seen, was not recognized by the authors of [Bacry and Levy-Leblond (1968)], and this fact is reflected in the names and notations of these groups. Similarly, there are no indications of this interpretation in the papers [Fernandez Sanjuan (1984)], [Rembielinski and Tybor (1984)]. Further Carroll kinematics will be denoted as $\mathcal{C}_4(j_1)$, and their kinematic groups as $G(j_1) = SO(5; j_1, 1, 1, \iota_4)$.

Bacry and Levy-Leblond (1968) have described 11 kinematic groups. Nine of them have obtained geometrical interpretation as spaces of constant curvature. The remaining two kinematics — "para-Galilean" and static — cannot be identified with any of the spaces of constant curvature.

For example, kinematic "para-Galilean" group is obtained from Galilean group $SO(5; \iota_1, \iota_2)$ by substitution $\mathbf{P} \to \mathbf{K}$, $\mathbf{K} \to \mathbf{P}$, i.e. under the new interpretation of generators, in which the generators of spatial translations of Galilean kinematics are claimed to be the generators of boosts of "para-Galilean" kinematics, and the generators of Galilean boosts — to be the generators of spatial "para-Galilean" translations.

2.2 Carroll kinematics

Let us describe Carroll kinematics $C_4(j_1)$ in detail. Each of the kinematics $C_4(j_1)$ is a trivial fiber space, the base of which is a three-dimensional subspace, interpreted as an absolute isotropic physical space, and the fiber is a one-dimensional subspace, interpreted as time. In comparison with Galilean kinematics, space and time in Carroll kinematics seem as if their properties were interchanged. In contrast with the mathematical features of Carroll kinematics, time is the base and the isotropic physical space is the fiber in Galilean kinematics. From the physical point of view, time is absolute in Galilean kinematics where two simultaneous events in the frame K are simultaneous in any other frame obtained by Galilean transformation (boost) of K; on the other hand, space is absolute in Carroll kinematics, i.e. two events, happened in the same spatial point of the frame K (unispatial events), happen in the same point of space (remain unispatial) in any other frame obtained from K by transformation of boost.

The method of transitions between groups enables us to easily obtain Casimir operators of Lie algebra of Carroll group $G(j_1)$ from the known invariant operators of algebra $so(5)$

$$I_1(j_1) = H^2 + j_1^2 \mathbf{K}^2, \quad I_2(j_1) = (H\mathbf{J} - \mathbf{P} \times \mathbf{K})^2 + j_1^2 (\mathbf{K}, \mathbf{J})^2. \quad (2.5)$$

For $j_1 = \iota_1$ the operators (2.5) coincide with Casimir operators of Carroll group, obtained by Levy-Leblond (1965).

Carroll algebra is the semidirect sum

$$AG_4(j_1) = A_4 \oplus M_6, \quad A_4 = \{H, \mathbf{K}\}, \quad M_6 = \{\mathbf{P}, \mathbf{J}\}; \quad (2.6)$$

besides (2.6), it allows the expansion

$$AG_4(j_1) = A_4' \oplus (A_3 \oplus M_3), \quad A_4' = \{H, \mathbf{P}\}, \quad A_3 = \{\mathbf{K}\}, \quad M_3 = \{\mathbf{J}\}. \quad (2.7)$$

Let us introduce two types of special expansions of the finite group transformations of group $G(j_1)$:

$$g = T^{\mathbf{r}} T^t T^{\mathbf{v}} R \equiv (\mathbf{r}, t, \mathbf{v}, R), \tag{2.8}$$

$$g = T^l T^{\mathbf{u}} T^{\mathbf{r}} R \equiv (l, \mathbf{u}, \mathbf{r}, R), \tag{2.9}$$

where $R = \exp(\mathbf{wJ})$ is the spatial rotation; $T^{\mathbf{v}} = \exp(\mathbf{vK})$ is the boost; $T^{\mathbf{r}} = \exp(\mathbf{rP})$ is the spatial translation; $T^t = \exp(tH)$ is the temporal translation. The structure of the semidirect product for groups $G(j_1)$ is determined by the expansions of their algebras (2.6), (2.7) and can be written as

$$G(j_1) = (T^t \times T^{\mathbf{v}}) \otimes (T^{\mathbf{r}} \times R), \quad G(j_1) = (T^t \times T^{\mathbf{r}}) \otimes (T^{\mathbf{v}} \times R). \tag{2.10}$$

To find algebraically the group multiplication rule for the elements of group $G(j_1)$, written as a special expansion (2.8), let us apply the theorem on the exact representations and consider the adjoint representation $(C_z)_{m_1 m_2} = \sum_m c^{m_1}_{m m_2} z^m$ of algebra $AG(j_1)$, which is exact, because the center of the algebra is zero. Having ordered the generators $H, \mathbf{P}, \mathbf{K}, \mathbf{J}$, we write down the matrix of adjoint representation, corresponding to canonical parameters $\tilde{t}, \tilde{r}_k, \tilde{v}_k, \tilde{w}_k$ (so that the general element $a \in AG(j_1)$ is $a = \tilde{t}H + \tilde{\mathbf{r}}\mathbf{P} + \tilde{\mathbf{v}}\mathbf{K} + \tilde{\mathbf{w}}\mathbf{J}$),

$$C_z = \begin{pmatrix} 0 & -\tilde{\mathbf{v}} & \tilde{\mathbf{r}} & 0 \\ 0 & -A(\tilde{\mathbf{w}}) & 0 & -A(\tilde{\mathbf{r}}) \\ -j_1^2 \tilde{\mathbf{r}} & j_1^2 \tilde{t} \mathbf{1} & -A(\tilde{\mathbf{w}}) & -A(\tilde{\mathbf{v}}) \\ 0 & -j_1^2 A(\tilde{\mathbf{r}}) & 0 & -A(\tilde{\mathbf{w}}) \end{pmatrix}, \tag{2.11}$$

where $\tilde{\mathbf{v}}, \tilde{\mathbf{r}}$ are vector-rows; $j_1^2 \tilde{\mathbf{r}}$ is a vector-column; $A(\mathbf{x})$ is the matrix of adjoint representation of the rotation group

$$A(\mathbf{x}) = \begin{pmatrix} 0 & x_3 & -x_2 \\ -x_3 & 0 & x_1 \\ x_2 & -x_1 & 0 \end{pmatrix}. \tag{2.12}$$

The matrices of the adjoint representation, corresponding to the special expansions, can be derived from (2.11). Really, let us put $\tilde{\mathbf{r}} = \tilde{\mathbf{v}} = \tilde{\mathbf{w}} = 0$, then $\tilde{t} = t$, i.e. canonical parameter $\tilde{t}$ coincides with the special parameter t, and formula (2.11) gives the expression for the matrix C_t. Putting $\tilde{t} = \tilde{\mathbf{r}} = \tilde{\mathbf{w}} = 0$ in (2.11), we get $\tilde{\mathbf{v}} = \mathbf{v}$ and find the matrix $C_{\mathbf{v}}$. For $\tilde{t} = \tilde{\mathbf{v}} = \tilde{\mathbf{w}} = 0$

we get $\tilde{\mathbf{r}} = \mathbf{r}$ and the matrix $C_{\mathbf{r}}$. At last, for $\tilde{t} = \tilde{\mathbf{r}} = \tilde{\mathbf{v}} = 0$, i.e. for $\tilde{\mathbf{w}} = \mathbf{w}$ the formula (2.11) gives the matrix $C_{\mathbf{w}}$.

Using the theorem on exact representations, we obtain convenient relations

$$RT^{\mathbf{r}}R^{-1} = T^{R\mathbf{r}}, \quad RT^{\mathbf{v}}R^{-1} = T^{R\mathbf{v}},$$

$$R = \exp A(\mathbf{w}) = 1 + A(\mathbf{w})\frac{\sin w}{w} + A^2(\mathbf{w})\frac{1 - \cos w}{w^2},$$

$$A^2(\mathbf{w}) = \begin{pmatrix} -w_2^2 - w_3^2 & w_1 w_2 & w_1 w_3 \\ w_1 w_2 & -w_1^2 - w_3^2 & w_2 w_3 \\ w_1 w_3 & w_2 w_3 & -w_1^2 - w_2^2 \end{pmatrix},$$

$$T^t T^{\mathbf{r}} = T^{\mathbf{r}} T^{t_1} T^{\mathbf{v}_1},$$

$$t_1 = t \cos j_1 r, \quad \mathbf{v}_1 = j_1^2 t \mathbf{r} \frac{\sin j_1 r}{j_1 r},$$

$$T^{\mathbf{u}} T^{\mathbf{r}} = T^{\mathbf{r}} T^{\mathbf{u}_1} T^{t_1},$$

$$\mathbf{u}_1 = \mathbf{u} - (\mathbf{r}, \mathbf{u})\mathbf{r}\frac{1 - \cos j_1 r}{r^2}, \quad t_1 = -(\mathbf{r}, \mathbf{u})\frac{\sin j_1 r}{j_1 r}. \tag{2.13}$$

Here $r = (r_1^2 + r_2^2 + r_3^2)^{1/2}$. The temporal translations and boosts commute $T^t T^{\mathbf{v}} = T^{\mathbf{v}} T^t$. For the spatial translations we get $T^{\mathbf{r}} T^{\mathbf{r}_1} = T^{\mathbf{r}'}$, where vector $\mathbf{r}' = \mathbf{r} \oplus \mathbf{r}_1$ — the generalized sum of vectors $\mathbf{r}$ and $\mathbf{r}_1$ — can be found from the relations

$$\cos j_1 r' = \cos j_1 r \cos j_1 r_1 - j_1^2 (\mathbf{r}, \mathbf{r}_1) \frac{\sin j_1 r \sin j_1 r_1}{j_1 r j_1 r_1},$$

$$\mathbf{r}' \frac{\sin j_1 r'}{j_1 r'} = \mathbf{r}' \frac{\sin j_1 r}{j_1 r} + \mathbf{r}_1 \cos j_1 r \frac{\sin j_1 r_1}{j_1 r_1}$$

$$- \mathbf{r}_1 (\mathbf{r}, \mathbf{r}_1) \frac{1 - \cos j_1 r_1}{r_1^2} \frac{\sin j_1 r}{j_1 r}. \tag{2.14}$$

The sum $\mathbf{r} \oplus \mathbf{r}_1$ for $j_1 = 1$ is the addition rule for vectors of the three-dimensional spherical space, for $j_1 = \iota_1$ is the usual vector addition in the three-dimensional Euclidean space, and for $j_1 = i$ is the addition rule for vectors in the three-dimensional Lobachevsky space ([Beresin, Kurochkin and Tolkachev (1989)]), because such is the geometry of the base of fibering for the corresponding values of the parameter j_1.

Now it is easy to find the multiplication rule for the elements of group $G(j_1)$, and for the special expansion (2.8) it is as follows:

$$(\mathbf{r}, t, \mathbf{v}, R)(\mathbf{r}_1, t_1, \mathbf{v}_1, R_1) = (\mathbf{r} \oplus R\mathbf{r}_1, t', \mathbf{v}', RR_1),$$

$$t' = t_1 + t \cos j_1 r_1 - (\mathbf{v}, R\mathbf{r}_1)\frac{\sin j_1 r_1}{j_1 r_1},$$

$$\mathbf{v}' = \mathbf{v} + R\mathbf{v}_1 + R\mathbf{r}_1\left[j_1^2 t \frac{\sin j_1 r}{j_1 r} - (\mathbf{v}, R\mathbf{r}_1)\frac{1 - \cos j_1 r_1}{r_1^{-2}}\right]. \qquad (2.15)$$

The inverse element can be presented as

$$(\mathbf{r}, t, \mathbf{v}, R)^{-1} = (-R^{-1}\mathbf{r}, t'', \mathbf{v}'', R^{-1}),$$

$$t'' = -t \cos j_1 r - (\mathbf{v}, \mathbf{r})\frac{\sin j_1 r}{j_1 r},$$

$$\mathbf{v}'' = -R^{-1}\left\{\mathbf{v} - \mathbf{r}\left[j_1^2 t \frac{\sin j_1 r}{j_1 r} + (\mathbf{v}, \mathbf{r})\frac{1 - \cos j_1 r}{r^2}\right]\right\}. \qquad (2.16)$$

We also write down the composition law for the expansion (2.9)

$$(l, \mathbf{u}, \mathbf{r}, R)(l_1, \mathbf{u}_1, \mathbf{r}_1, R_1) = (l', \mathbf{u}', \mathbf{r} \oplus R\mathbf{r}_1, RR_1),$$

$$l' = l + l_1 \cos j_1 r + (\mathbf{r}, R\mathbf{u}_1)\frac{\sin j_1 r}{j_1 r},$$

$$\mathbf{u}' = \mathbf{u} + R\mathbf{u}_1 - \mathbf{r}\left\{j_1^2 l_1 \frac{\sin j_1 r}{j_1 r} + (\mathbf{r}, R\mathbf{u}_1)\frac{1 - \cos j_1 r}{r^2}\right\}. \qquad (2.17)$$

The inverse element is as follows

$$(l, \mathbf{u}, \mathbf{r}, R)^{-1} = (l'', \mathbf{u}'', -R^{-1}\mathbf{r}, R^{-1}),$$

$$l'' = -l \cos j_1 r + (\mathbf{r}, \mathbf{u})\frac{\sin j_1 r}{j_1 r},$$

$$\mathbf{u}'' = R^{-1}\left\{\mathbf{u} + \mathbf{r}\left[-j_1^2 l \frac{\sin j_1 r}{j_1 r} + (\mathbf{r}, \mathbf{u})\frac{1 - \cos j_1 r}{r^2}\right]\right\}. \qquad (2.18)$$

Carroll kinematics can be obtained as factor-space $\mathcal{C}_4(j_1) = G(j_1)/\{T^{\mathbf{v}} \times R\}$, with the spatial-temporal coordinates, given by the transformations $T^{\mathbf{r}}T^t = T^l T^{\mathbf{r}}$. The transformation $T^{\mathbf{r}}T^t$ determines the so-called quasi-Beltramian (or equidistant) coordinates $(\mathbf{r}, t)$ of the kinematics point M, and $T^l T^{\mathbf{r}}$ — the Cartesian coordinates $(\mathbf{r}, l)$ of the point M (Fig. 2.1), and, what is more, the spatial coordinates of the point M are the same,

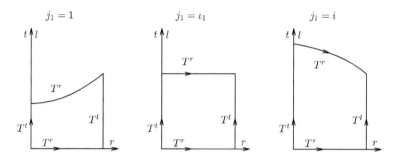

Fig. 2.1   Cartesian $(l, \mathbf{r})$ and quasi-Beltramian (equidistant) $(\mathbf{r}, \mathbf{l})$ coordinates in Carroll kinematics.

and the temporal coordinates are related by the formula $t = l \cos j_1 r$. In the Carroll kinematics $C_4(\iota_1)$ both transformations determine the same Cartesian coordinate system.

Transitive action of Carroll group in kinematics $C_4(j_1)$, corresponding to the group element $(\mathbf{a}, b, \mathbf{v}, R)$, can be written in the coordinates $(\mathbf{r}, t)$ as follows:

$$t' = t + b \cos j_1 r - (\mathbf{v}, R\mathbf{r}) \frac{\sin j_1 r}{j_1 r}, \quad \mathbf{r}' = \mathbf{a} \oplus R\mathbf{r}. \qquad (2.19)$$

The generators of Carroll group make a direct physical sense by being observable variables for a free physical system (particle) in Carroll kinematics. For this reason the physical sense of numerical values of the generators in some state $\langle s|$ is given by energy $h = \langle s|H \rangle$, momentum $\mathbf{p} = \langle s|\mathbf{P} \rangle$, location of mass center $\mathbf{k} = \langle s|\mathbf{K} \rangle$ and angular momentum $\mathbf{l} = \langle s|\mathbf{J} \rangle$ of a free system. Under general Carroll transformation $(\mathbf{a}, b, \mathbf{v}, R)$ the numerical values of generators, shown in monograph [Zaitsev (1974)], are transformed according to the adjoint representation of the group:

$$(h', \mathbf{p}', \mathbf{k}', \mathbf{l}') = (h, \mathbf{p}, \mathbf{k}, \mathbf{l}) \exp C_{\mathbf{w}} \exp C_{\mathbf{v}} \exp C_b \exp C_{\mathbf{r}'}, \qquad (2.20)$$

where a row of numerical values is multiplied by a row of matrix of adjoint representation. Taking into account that $\exp C_t = 1 + C_t$, $\exp C_{\mathbf{v}} = 1 + C_{\mathbf{v}}$,

$$\exp C_{\mathbf{a}} = 1 + C_{\mathbf{a}} \frac{\sin j_1 a}{j_1 a} + C_{\mathbf{a}}^2 \frac{1 - \cos j_1 a}{(j_1 a)^2}, \qquad (2.21)$$

we derive the transformation rules for the energy, momentum, location of mass center and angular momentum of a free physical system in Carroll kinematics from (2.20):

$$h' = h\cos j_1 a + j_1^2(\mathbf{a}, R\mathbf{k})\frac{\sin j_1 a}{j_1 a},$$

$$\mathbf{p}' = R\mathbf{p}\cos j_1 a + \mathbf{v}\left[h\cos j_1 a + j_1^2(\mathbf{a}, R\mathbf{k})\frac{\sin j_1 a}{j_1 a}\right]$$

$$- j_1^2 R\mathbf{k}\left[b\cos j_1 a - (\mathbf{a}, \mathbf{v})\frac{\sin j_1 a}{j_1 a}\right]$$

$$+ \mathbf{a}\frac{1 - \cos j_1 a}{a^2}\left[(\mathbf{a}, R\mathbf{p}) + h(\mathbf{a}, \mathbf{v}) - j_1^2 b(\mathbf{a}, R\mathbf{k})\right],$$

$$\mathbf{k}' = R\mathbf{k} - \mathbf{a}(\mathbf{a}, R\mathbf{k})\frac{1 - \cos j_1 a}{a^2} - \mathbf{a}h\frac{\sin j_1 a}{j_1 a},$$

$$\mathbf{l}' = R\mathbf{l}\cos j_1 a - \mathbf{v}\times R\mathbf{k}\cos j_1 a - \mathbf{a}\times R\mathbf{l}\frac{\sin j_1 a}{j_1 a}$$

$$+ \mathbf{a}\times(\mathbf{v}\times R\mathbf{k})\frac{\sin j_1 a}{j_1 a} + \mathbf{a}\frac{1 - \cos j_1 a}{a^2}\left[(\mathbf{a}, R\mathbf{l}) - (\mathbf{a}, \mathbf{v}\times R\mathbf{k})\right]. \quad (2.22)$$

The state of a free system can be characterized by eigenvalues of Casimir operators (2.5). It is easy to check that under the transformations (2.22) two quantities remain invariant:

$$\text{inv}_1 = h^2 + j_2^2\mathbf{k}^2, \quad \text{inv}_2 = (h\mathbf{l} - \mathbf{p}\times\mathbf{k})^2 + j_1^2(\mathbf{k},\mathbf{l})^2. \quad (2.23)$$

2.3 Non-relativistic kinematics

Semispherical group $SO(5; 1, \iota_2)$ and semihyperbolic (or Newton) group $SO(5; i, \iota_2)$ are groups of motions for non-relativistic kinematics, but, in contrast to Galilean kinematics, they have constant curvature, positive and negative, correspondingly. The representations of these groups, as well as the description of physical systems in such kinematics, are given in [Derom and Dubois (1972); Dubois (1973)]. For the sake of completeness, we shall briefly describe non-relativistic kinematics in this section, including Galilean kinematics with group of motions $SO(5; \iota_1, \iota_2)$.

The commutators of the generators of group $SO(5; j_1, \iota_2)$, $j_1 = 1, \iota_1, i$ are given by (2.3) for $j_2 = \iota_2$, from which it follows immediately that algebra $so(5; j_1, \iota_2)$ can be presented as a semidirect sum

$$so(5; j_1, \iota_2) = L_6 \niplus L_4, \quad L_6 = \{\mathbf{K}, \mathbf{J}\}, \quad L_4 = \{h, \mathbf{P}\}. \quad (2.24)$$

Let us consider special expansion of the finite transformations from $SO(5; j_1, \iota_2)$

$$g = T^t T^{\mathbf{r}} T^{\mathbf{u}} T^{\mathbf{w}} \equiv (t, \mathbf{r}, \mathbf{u}, \mathbf{w}). \qquad (2.25)$$

The definition and the physical sense of the factors are the same as in the previous section. The matrix of the adjoint representation, corresponding to the canonical parameters $\tilde{t}, \tilde{\mathbf{r}}, \tilde{\mathbf{u}}, \tilde{\mathbf{w}}$, is as follows:

$$C = \begin{pmatrix} 0 & 0 & 0 & 0 \\ -\tilde{\mathbf{u}} & A(\tilde{\mathbf{w}}) & -\tilde{t}\mathbf{1} & A(\tilde{\mathbf{r}}) \\ 0 & j_1^2 \tilde{t}\mathbf{1} & A(\tilde{\mathbf{w}}) & A(\tilde{\mathbf{u}}) \\ 0 & 0 & 0 & A(\tilde{\mathbf{w}}) \end{pmatrix}, \qquad (2.26)$$

where $A(\mathbf{x})$ is given by (2.12). As in the previous section, we get

$$T^{\mathbf{w}} T^{\mathbf{r}} = T^{R\mathbf{r}} T^{\mathbf{w}}, \quad T^{\mathbf{w}} T^{\mathbf{u}} = T^{R\mathbf{u}} T^{\mathbf{w}}, \quad T^{\mathbf{u}} T^{\mathbf{r}} = T^{\mathbf{r}} T^{\mathbf{u}},$$

$$T^t T^{\mathbf{r}} = T^{\mathbf{r}_1} T^{\mathbf{u}_1} T^t, \quad \mathbf{r}_1 = \mathbf{r} \cos j_1 t, \quad \mathbf{u}_1 = \mathbf{r} j_1 \sin j_1 t,$$

$$T^t T^{\mathbf{u}} = T^{\mathbf{r}_1} T^{\mathbf{u}_1} T^t, \quad \mathbf{r}_1 = -\mathbf{u}_1 j_1^{-1} \sin j_1 t, \quad \mathbf{u}_1 = \mathbf{u} \cos j_1 t, \quad (2.27)$$

which enables us to derive the multiplication rule for the elements of the group $SO(5; j_1, \iota_2)$

$$(t, \mathbf{r}, \mathbf{u}, R) \cdot (t_1, \mathbf{r}_1, \mathbf{u}_1, R_1) = (t + t_1, \mathbf{r}', \mathbf{u}', RR_1),$$

$$\mathbf{r}' = R\mathbf{r}_1 + \mathbf{r} \cos j_1 t + \mathbf{u} j_1^{-1} \sin j_1 t,$$

$$\mathbf{u}' = R\mathbf{u}_1 + \mathbf{u} \cos j_1 t - \mathbf{r} j_1 \sin j_1 t. \qquad (2.28)$$

The inverse element is as follows

$$(t, \mathbf{r}, \mathbf{u}, R)^{-1} = (-t, \mathbf{r}'', \mathbf{u}'', R^{-1}),$$

$$\mathbf{r}'' = -R^{-1}(\mathbf{r} \cos j_1 t - \mathbf{u} j_1^{-1} \sin j_1 t),$$

$$\mathbf{u}'' = -R^{-1}(\mathbf{u} \cos j_1 t + \mathbf{r} j_1 \sin j_1 t). \qquad (2.29)$$

The numerical values of the generators in some state are transformed according to the adjoint representation. For the special expansion (2.25) this representation is as follows:

$$h' = h - (\mathbf{u}, R\mathbf{p}) + j_1^2(\mathbf{r}, R\mathbf{l}), \quad \mathbf{p}' = R\mathbf{p} \cos j_1 t + R\mathbf{k} j_1 \sin j_1 t,$$

$$\mathbf{k}' = R\mathbf{k} \cos j_1 t - R\mathbf{p} \frac{1}{j_1} \sin j_1 t, \quad \mathbf{l}' = R\mathbf{l} + \mathbf{u} \times R\mathbf{k}. \qquad (2.30)$$

Here h is energy; $\mathbf{p}$ is momentum; $\mathbf{l}$ is angular momentum of a free, zero mass particle in non-relativistic kinematics, and vector $\mathbf{k}$ is proportional to its location.

Groups $SO(5; j_1, \iota_2)$ are the only ones among the kinematic groups which pose as ray representation. The latter can be characterized by one number $m \in \mathbb{R}$. Correspondingly, the algebras $so(5; j_1, \iota_2)$ allow nontrivial central extension, which can be obtained by adding to ten generators the identity operator I and changing for $j_2 = \iota_2$ the commutator $[P_l, K_l] = 0$ to the commutator $[P_l, K_l] = mI$ in (2.3). Matrix of the adjoint representation for the central extension can be derived from (2.26) by adding the last zero column and the last row of the form $(0, -m\tilde{\mathbf{u}}, m\tilde{\mathbf{r}}, 0, 0)$. Then, taking into account that the numerical value of the unit operator is constant, which can be considered to be equal to a unit, instead of (2.30) we get the formulas for transformations of energy, momentum, angular momentum and location of a free particle with mass m

$$h' = h - (\mathbf{u}, R\mathbf{p}) + j_1^2(\mathbf{r}, R\mathbf{k}) + \frac{mu^2}{2} + j_1^2 \frac{mr^2}{2},$$

$$\mathbf{p}' = (R\mathbf{p} - m\mathbf{u})\cos j_1 t + (R\mathbf{k} + m\mathbf{r})j_1 \sin j_1 t,$$

$$\mathbf{k}' = (R\mathbf{k} + m\mathbf{r})\cos j_1 t - (R\mathbf{p} - m\mathbf{u})\frac{1}{j_1}\sin j_1 t,$$

$$\mathbf{l}' = R\mathbf{l} + u \times R\mathbf{k} + \mathbf{r} \times R\mathbf{p}. \tag{2.31}$$

Casimir operators of the algebra $so(5; j_1, \iota_2)$ are given by (1.49). Putting $j_2 = \iota_2$, $j_3 = j_4 = 1$, $n = 4$, we get

$$C_2(j_1) = \mathbf{P}^2 + j_1^2 \mathbf{K}^2, \quad C_4(j_1) = (\mathbf{P} \times \mathbf{K})^2. \tag{2.32}$$

The physical sense of the operator $C_2(j_1)$ is given by the kinetic energy of a zero mass particle, and the operator $C_4(j_1)$ is a squared angular momentum of this particle. Under the central extension we get three Casimir operators

$$C_1' = mI, \quad C_2' = H^2 - \frac{1}{2m}(\mathbf{P}^2 + j_1^2\mathbf{K}), \quad C_4' = \left(\mathbf{J} - \frac{1}{m}(\mathbf{K} \times \mathbf{P})\right)^2, \tag{2.33}$$

where C_1' is a mass operator of a particle; C_2' is an operator of the particle's internal energy, equal to the difference between total and kinetic energy, and C_4' is associated to the squared difference between the total momentum of a particle and its angular momentum, i.e. the squared internal momentum (spin) operator of a particle. So in the group approach the notion of a spin for a classical particle arises quite naturally, both in Galilean kinematics and in Newton kinematics.

As it has been established by V. Bargmann (1954):

Definition 2.1. The ray unitary representation U of group G is given by

$$U(g)U(g_1) = U(gg_1)\exp\{if(g, g_1)\}, \quad g, g_1 \in G, \tag{2.34}$$

where the real-valued function $f(g, g_1)$ is called a multiplicator, has a property $f(1, g) = f(g, 1) = 0$ and satisfies the condition implied by Jacobi identity

$$f(g, g_1) + f(gg_1, g_2) = f(g, g_1 g_2) + f(g_1, g_2). \tag{2.35}$$

Multiplying each operator of the representation $U(g)$ by the continuous function $\exp\{iv(g)\}$, we come to a new ray representation by operators $U'(g) = U(g)\exp\{iv(g)\}$ having the multiplicator

$$f'(g, g_1) = f(g, g_1) + v(g) + v(g_1) - v(gg_1). \tag{2.36}$$

J.-G. Dubois (1973) had shown that groups $SO(5; j_1, \iota_2)$ have a ray representation with the multiplicator

$$f'(g, g_1) = m\left[(\mathbf{u}^2 - j_1^2\mathbf{r}^2)\frac{\sin 2j_1 t_1}{4j_1} - (\mathbf{r}, \mathbf{u})\sin^2 j_1 t_1\right.$$

$$\left. - (\mathbf{r}, R\mathbf{r})j_1 \sin j_1 t_1 + (\mathbf{u}, R\mathbf{r}_1)\cos j_1 t_1\right], \tag{2.37}$$

where $g, g_1 \in SO(5; j_1, \iota_2)$ are written as expansions (2.25). The method, proposed by V. Bargmann (1954), brings to the multiplicator

$$f(g, g_1) = m[(\mathbf{r}, R\mathbf{r}_1)j_1 \sin j_1 t_1 + (\mathbf{r}, R\mathbf{u}_1)\cos j_1 t_1$$

$$- (\mathbf{u}, R\mathbf{r}_1)\cos j_1 t_1 + (\mathbf{u}, R\mathbf{u}_1)j_1^{-1}\sin j_1 t_1]/2. \tag{2.38}$$

The multiplicators (2.37) and (2.38) are connected via (2.36) with the function $v(g) = m(\mathbf{r}, \mathbf{u})$. For $j_1 = \iota_1$ the expression (2.37) coincides with the multiplicator of Galilean group, found by J. Voisin (1965).

Chapter 3

The Jordan–Schwinger representations of Cayley–Klein groups

In 1935, [P. Jordan (1935)] introduced the so-called Jordan mapping that is a mapping from a one-particle realization of the kinematic symmetry into field operators of either boson or fermion type. This mapping preserves the commutation relations of matrices. In 1952, [J. Schwinger (1965)] introduced an original treatment of the rotation group by representing the matrix generators in terms of their bilinear forms with respect to boson annihilation and creation operators. Since this representation is equivalent to the Jordan mapping it is often called [Kim (1987)] the Jordan–Schwinger representation. It has been widely used to provide a treatment of representations of Lie groups. On the other hand, there is the well-developed theory of a many-body quantum system in the second quantized field formalism whose Hamiltonians are multidimensional quadratic in boson or fermion creation and annihilation operators [Dodonov and Man'ko (1987); Malkin and Man'ko (1979)]. The methods of this theory may be used in calculating matrix elements of finite transformations of Lie groups in the bases of coherent and Fock states [Man'ko and Trifonov (1987)].

In the present chapter we shall discuss the Jordan–Schwinger representation of the orthogonal, special unitary, and symplectic Cayley–Klein groups [Gromov and Man'ko (1990)]. We consider the Jordan–Schwinger representation of the matrix generators of the groups under discussion based on either fermion or boson operators. For groups derived from classical groups only by contractions the set of particle operators describing the representation is pure, i.e., all members of the set are either annihilation or creation operators. However, if the groups are obtained by solely analytical continuations or both continuations and contractions, then the representations are based on mixed sets of annihilation and creation operators [Kim (1987)]. The matrix elements of the Jordan–Schwinger representation of the finite group transformations are calculated in the bases of coherent states, which were introduced by R.J. Glauber (1963). In the case of boson representations we use an important property of the coherent states, namely, that the coherent state gives a generating function for discrete Fock states. Then the matrix elements of the finite group transformations in the coherent state bases are the generating function for the matrix elements in Fock bases. The last matrix elements are expressed in terms of Hermite polynomials of several variables with zero arguments.

3.1   The second quantization method and matrix elements

Let $\mathbf{G}$ be a group of N-dimensional matrices with the generators X_k and commutators $[X_k, X_m] = \sum_s x^s_{km} X_s$. By definition, the operators $\hat{X}_k = \sum_{p,q} (X_k)_{pq} \hat{a}^+_p \hat{a}_q$, where $\hat{a}^+_p$ and $\hat{a}_q$ are the boson or fermion creation and annihilation operators, respectively, satisfying the canonical commutation relations

$$\hat{a}_i \hat{a}_k = \epsilon \hat{a}_k \hat{a}_i, \quad \hat{a}^+_i \hat{a}^+_k = \epsilon \hat{a}^+_k \hat{a}^+_i,$$
$$\hat{a}_i \hat{a}^+_k - \epsilon \hat{a}^+_k \hat{a}_i = \delta_{ik}. \tag{3.1}$$

Here we have $\epsilon = 1$ in the boson case and $\epsilon = -1$ in the fermion case. Then the operators $\hat{X}_k$ satisfy the commutation relations of the Lie algebra of group G and realize their Jordan–Schwinger representation. The finite group transformation operator $\hat{U}_g(\mathbf{r})$ is connected with the general element $\hat{X}(\mathbf{r}) = \sum_k r_k \hat{X}_k$ of the Lie algebra by the exponential map $\hat{U}_g(\mathbf{r}) = \exp(-\hat{X}(\mathbf{r}))$, where r_k are the group parameters.

The representation space H is the state space of the N-dimensional quantum oscillator. For the bases in the representation space, we shall use the overcomplete family of Glauber coherent states [Glauber (1963)], i.e., the eigenstates of the annihilation operators $\hat{a}_k|\alpha\rangle = \alpha_k|\alpha\rangle$, $\langle\alpha|\alpha\rangle = 1$. In the boson case, α_k, $k = 1, 2, \ldots, N$ are complex variables and in the fermion case, α_k are Grassmann anticommutative variables [Beresin (1966)]. A vector $|\mathbf{f}\rangle \in H$ is determined [Man'ko and Trifonov (1987)] by the analytic (with respect to α^*) function $f(\alpha^*)$

$$f(\alpha^*) = \exp\left(\frac{1}{2}|\alpha|^2\right) \langle\alpha|f\rangle, \tag{3.2}$$

and the operator $\hat{U}_g$ by the kernel $U(\alpha^*, \beta)$

$$U(\alpha^*, \beta) = \exp\left(\frac{1}{2}|\alpha|^2 + \frac{1}{2}|\beta|^2\right) \langle\alpha|\hat{U}_g|\beta\rangle. \tag{3.3}$$

A transformed vector $|\mathbf{f}'\rangle = \hat{U}_g|\mathbf{f}\rangle$ is represented by the function

$$f'(\alpha^*) = \int U(\alpha^*, \beta) f(\beta^*) d\mu(\beta),$$
$$d\mu(\beta) = \pi^{-N} \exp(-|\beta|^2) d^2\beta, \tag{3.4}$$

where $|\alpha|^2 = \sum_{k=1}^{N} |\alpha_k|^2$, $|\beta|^2 = \sum_{k=1}^{N} |\beta_k|^2$, $d^2\beta = \prod_{k=1}^{N} d(\mathrm{Re}\beta_k) \cdot d(\mathrm{Im}\beta_k)$.

The matrix elements (or the kernel) of the finite group transformation operator $\hat{U}_g(\mathbf{r})$ are obtained by the method of motion integrals [Malkin and Man'ko (1979)]. The motion invariants are built with the help of the matrix

$$\Lambda(\mathbf{r}) = \exp(\Sigma B(\mathbf{r})) \equiv \begin{pmatrix} \xi & \eta \\ \eta_1 & \xi_1 \end{pmatrix}, \tag{3.5}$$

where

$$\Sigma = \begin{pmatrix} 0 & E \\ -\epsilon E & 0 \end{pmatrix},$$

E is an N-dimensional unit matrix and the matrix $B(\mathbf{r})$ is defined by the equation

$$\hat{X}(\mathbf{r}) = \sum_k r_k \hat{X}_k = (\hat{\mathbf{a}}, \hat{\mathbf{a}}^+) B(\mathbf{r}) \begin{pmatrix} \hat{\mathbf{a}} \\ \hat{\mathbf{a}}^+ \end{pmatrix}. \tag{3.6}$$

Here $(\hat{\mathbf{a}}, \hat{\mathbf{a}}^+)$ is the row matrix, $\begin{pmatrix} \hat{\mathbf{a}} \\ \hat{\mathbf{a}}^+ \end{pmatrix}$ is the column matrix and the product in Eq. (3.6) is the ordinary matrix product. We shall use such an agreement throughout the chapter. The kernel of the operator $\hat{U}_g(\mathbf{r})$ is given by the following equation [Malkin and Man'ko (1979)]:

$$U(\alpha^*, \beta, \mathbf{r}) = (\det \xi)^{-\epsilon/2} \exp\left(-\frac{1}{2}(\alpha^*, \beta) R(\mathbf{r}) \begin{pmatrix} \alpha^* \\ \beta \end{pmatrix}\right)$$

$$= (\det \xi)^{-\epsilon/2} \exp(-\frac{1}{2}\alpha^*\xi^{-1}\eta\alpha^* + \alpha^*\xi^{-1}\beta + \frac{1}{2}\epsilon\beta\eta_1\xi^{-1}\beta), \tag{3.7}$$

where the $2N$-dimensional matrix $R(\mathbf{r})$ is as follows:

$$R(\mathbf{r}) = \begin{pmatrix} \xi^{-1}\eta & -\xi^{-1} \\ -\epsilon\xi^{-1T} & -\epsilon\eta_1\xi^{-1} \end{pmatrix}. \tag{3.8}$$

In the boson case ($\epsilon = 1$) we also regard the discrete Fock states basis in the representation space H. The Fock state $|\mathbf{n}\rangle$ is the eigenstate of the particle number operator $\hat{a}_k^+ \hat{a}_k |\mathbf{n}\rangle = n_k |\mathbf{n}\rangle$, $\mathbf{n} = (n_1, n_2, \ldots, n_N)$ and n_k are non-negative integer numbers. We may use the important property of the coherent states, namely, that the coherent state gives the generating function for the Fock states

$$|\alpha\rangle = \exp\left(-\frac{1}{2}|\alpha|^2\right) \sum_{\mathbf{n}=0}^{\infty} \frac{\alpha^{\mathbf{n}}}{(\mathbf{n}!)^{1/2}} |\mathbf{n}\rangle, \qquad (3.9)$$

where

$$\mathbf{n}! = \prod_{k=1}^{N} n_k!, \quad \alpha^{\mathbf{n}} = \prod_{k=1}^{N} \alpha_k^{n_k}$$

and it follows immediately that the kernel (3.7) is the generating function for the matrix elements of the operator $\hat{U}_g(\mathbf{r})$ in the Fock states basis

$$U(\alpha^*, \beta, r) = \sum_{\mathbf{m},\mathbf{n}=0}^{\infty} \frac{\alpha^{*\mathbf{m}} \beta^{\mathbf{n}}}{(\mathbf{m}!\mathbf{n}!)^{1/2}} \langle \mathbf{m} | \hat{U}_g(\mathbf{r}) | \mathbf{n} \rangle. \qquad (3.10)$$

Multidimensional Hermite polynomials $H_k^{(R)}(\mathbf{x})$ are defined by their generating function as follows [Bateman and Erdelyi (1953)]:

$$\exp\left(-\frac{1}{2}\mathbf{a}R\mathbf{a} + \mathbf{a}R\mathbf{x}\right) = \sum_{k=0}^{\infty} \frac{\mathbf{a}^k}{\mathbf{k}!} H_k^{(R)}(\mathbf{x}). \qquad (3.11)$$

Then the kernel (3.10) multiplied by $(\det \xi)^{1/2}$ is the generating function for the Hermite polynomials of $2N$ zero variables

$$(\det \xi)^{1/2} U(\alpha^*, \beta, \mathbf{r}) = \exp\left(-\frac{1}{2}(\alpha^*, \beta)R(\mathbf{r})\begin{pmatrix} \alpha^* \\ \beta \end{pmatrix}\right)$$

$$= \sum_{\mathbf{m},\mathbf{n}=0}^{\infty} \frac{\alpha^{*\mathbf{m}} \beta^{\mathbf{n}}}{\mathbf{m}!\mathbf{n}!} H_{\mathbf{m},\mathbf{n}}^{(R(r))}(0), \qquad (3.12)$$

where the matrix $R(\mathbf{r})$ is given by Eq. (3.8).

3.2 The rotation groups in Cayley–Klein spaces

It is well known in geometry [Yaglom, Rosenfeld and Yasinskaya (1964)], that there are 3^n n-dimensional real spaces of constant curvature.

R.I. Pimenov (1965) had given their unified axiomatic description and had built the transformations of the elliptic space into arbitrary space of constant curvature. In accordance with the Erlangen Program, due to F. Klein, each geometry is associated with a motion group. Then the transformations of the geometry induce the transformations of the related motion group. This idea was used to develop the method of transitions between groups [Gromov (1981, 1990)] that naturally unify both contractions and analytical continuations of groups.

Let us use the fundamental map (1.43) of the Euclidean space $\mathbf{R}_{n+1}$ into the space $\mathbf{R}_{n+1}(j)$

$$\psi : \mathbf{R}_{n+1} \to \mathbf{R}_{n+1}(j),$$
$$\psi x_0^* = x_0, \quad \psi x_k^* = x_k(0, k), \tag{3.13}$$

where $k = 1, 2, \ldots, n; x_0, x_k$ are the Cartesian coordinates, $j = (j_1, j_2, \ldots, j_n)$, parameter j_k may be equal to the real unit 1, or to the Clifford dual unit ι_k, or to the imaginary unit i, symbol $(0, k)$ is given by Eq. (1.27). The $(n + 1)$-dimensional real Cayley–Klein space $\mathbf{R}_{n+1}(j)$ is defined as the $(n + 1)$-dimensional vector space with the following metric:

$$\mathbf{x}^2(j) = x_0^2 + \sum_{k=1}^{n} x_k^2(0, k)^2. \tag{3.14}$$

Then Eq. (3.13) is the mapping of the Euclidean space into the Cayley–Klein spaces. The space of constant curvature $\mathbf{S}_n(j)$ is realized on the sphere $\mathbf{S}_n(j) = \{\mathbf{x} \in R_{n+1}(\mathbf{j}) | \ \mathbf{x}^2(j) = 1\}$ in the Cayley–Klein space. The set of parameters j gives all 3^n $(n + 1)$-dimensional Cayley–Klein spaces or n-dimensional spaces of constant curvature.

The rotations of the Cayley–Klein space $\mathbf{R}_{n+1}(j)$ form the group $SO_{n+1}(j)$, which we call the orthogonal Cayley–Klein group. The map (3.13) induces the transformation of the group SO_{n+1} into the Cayley–Klein group $SO_{n+1}(j)$. The generators $\tilde{X}_{\mu\nu}$ of SO_{n+1} are the infinitesimal rotations in two-dimensional planes $\{x_\mu, x_\nu\}$, $\mu = 0, 1, \ldots, n - 1$, $\nu = 1, 2, \ldots, n, \mu < \nu$. The nonzero elements of the matrix $\tilde{X}_{\mu\nu}$ are as follows: $(\tilde{X}_{\mu\nu})_{\mu\nu} = -1, (\tilde{X}_{\mu\nu})_{\nu\mu} = 1$. It is easy to obtain the induced transformation law of the generators of SO_{n+1} under the map (3.13) in the following way [Gromov (1990)]:

$$X_{\mu\nu}(j) = (\mu, \nu)\hat{X}_{\mu\nu}(\to). \tag{3.15}$$

Here, by $\tilde{X}_{\mu\nu}(\rightarrow)$, we denote the transformed generator $\tilde{X}_{\mu\nu}$ with the following nonzero matrix elements:

$$(\tilde{X}_{\mu\nu}(\rightarrow))_{\mu\nu} = (\mu,\nu)(\tilde{X}_{\mu\nu})_{\mu\nu} = -(\mu,\nu),$$
$$(\tilde{X}_{\mu\nu}(\rightarrow))_{\nu\mu} = (\mu,\nu)^{-1}(\tilde{X}_{\mu\nu})_{\nu\mu} = (\mu,\nu)^{-1}. \tag{3.16}$$

Then the transformation (3.15) gives for the nonzero matrix elements of the generator $X_{\mu\nu}(j)$

$$(X_{\mu\nu}(j))_{\mu\nu} = -(\mu,\nu)^2, \quad (X_{\mu\nu})_{\nu\mu} = 1, \ \mu < \nu. \tag{3.17}$$

The generators $X_{\mu\nu}(j)$ satisfy the commutation relations (1.45)

$$[X_{\mu_1\nu_1}, X_{\mu_2\nu_2}] = \begin{cases} X_{\nu_1\nu_2}(\mu_1,\nu_1)^2, & \mu_1 = \mu_2, \quad \nu_1 < \nu_2 \\ X_{\mu_1\mu_2}(\mu_2,\nu_2)^2, & \mu_1 < \mu_2, \quad \nu_1 = \nu_2 \\ -X_{\mu_1\nu_2}, & \mu_1 < \nu_1 = \mu_2 < \nu_2 \end{cases} \tag{3.18}$$

of the group $SO_{n+1}(j)$.

Let us observe that when some parameters j_m are equal to the dual units the transformations (3.15) are the multidimensional Wigner–Inönü contractions. Indeed, $\tilde{X}_{\mu\nu}(\rightarrow)$ are the singularly transformed generators, the products (μ,ν) play the role of the zero tending parameters and the resulting generators (3.17) are not singular.

The finite rotation $\Xi(\mathbf{r},j) = \exp X(\mathbf{r},j)$ corresponds to the general element

$$X(\mathbf{r},j) = \sum_{\lambda=1}^{n(n+1)/2} r_\lambda X_\lambda(j), \quad r_\lambda \in \mathbb{R} \tag{3.19}$$

of the algebra $so_{n+1}(j)$. Here λ is in a one-to-one accordance with μ,ν, $\mu < \nu$ due to the equation

$$\lambda = \nu + \mu(n-1) - \mu(\mu-1)/2. \tag{3.20}$$

Due to the Cayley-Hamilton theorem [Korn and Korn (1961)] the matrix $\Xi(\mathbf{r},j)$ is expressed algebraically by the matrices $X^m(\mathbf{r},j)$, $m = 0,1,\ldots,n$. The explicit form of the finite rotations $\Xi(\mathbf{r},j)$ can be directly obtained for the groups of low dimensions, namely $SO_2(j_1), SO_3(j), SO_4(j)$.

Combining Eqs. (3.15) and (3.19) we observe that for the imaginary values of some parameters j_k some real group parameters r_λ are imaginary ones, i.e., they are analytically continued from the real number field into the complex one. The orthogonal group SO_{n+1} is transformed by these into some pseudoorthogonal group $SO(p,q)$. For the dual values of some parameters j_k some real group parameters r_λ are pure dual ones, i.e., they are continued into the dual number field and we have a contraction of the group SO_{n+1}. Thus, from the viewpoint of transformations, both procedures have the same nature: namely, the continuation of group parameters from the real number field into the dual (contraction) or complex ones.

3.3 The Jordan–Schwinger representations of the orthogonal Cayley–Klein groups

Let us define the transformation of annihilation and creation operators induced by the map (3.13) as follows:

$$\psi\hat{\mathbf{a}} = \left(\hat{a}_0, \hat{a}_k(0,k)\right), \quad \psi\hat{\mathbf{a}}^+ = \left(\hat{a}_0^+, \hat{a}_k^+(0,k)^{-1}\right), \quad k = 1, 2, \ldots, n, \tag{3.21}$$

where ψ is identical, i.e., $\psi\hat{\mathbf{a}} = \hat{\mathbf{a}}, \psi\hat{\mathbf{a}}^+ = \hat{\mathbf{a}}^+$, when $j_k = 1, \iota_k$. For the imaginary values of parameters j we use the well known properties of the annihilation and creation operators: $i\hat{a}_k = \hat{a}_k^+, i\hat{a}_k^+ = \epsilon\hat{a}_k$. Then Eq. (3.21) may be written in the form

$$\begin{pmatrix} \psi\hat{\mathbf{a}} \\ \psi\hat{\mathbf{a}}^+ \end{pmatrix} = \begin{pmatrix} \psi_1(j) & -\psi_2(j) \\ \epsilon\psi_2(j) & \psi_1(j) \end{pmatrix} \begin{pmatrix} \hat{\mathbf{a}} \\ \hat{\mathbf{a}}^+ \end{pmatrix} \equiv \Psi^{-1}(j) \begin{pmatrix} \hat{\mathbf{a}} \\ \hat{\mathbf{a}}^+ \end{pmatrix}. \tag{3.22}$$

where $\psi_1(j), \psi_2(j)$ are $(n+1)$-dimensional diagonal matrices with the following nonzero matrix elements: $(\psi_1(j))_{00} = 1, (\psi_1(j))_{kk} = \pm 1$, if $(0,k) = \pm b$ and b is a positive real or dual number; otherwise, $(\psi_1(j))_{kk} = 0(\psi_2(j))_{00} = 0$. $(\psi_2(j))_{kk} = 0$, if $(\psi_1(j))_{kk} = \pm 1$, and $(\psi_2(j))_{kk} = \mp 1$, if $(0,k) = \pm ib$. The $2(n+1)$-dimensional matrix $\Psi(j)$ has the property $\Psi(j) = (\Psi^{-1}(j))^T$. It is easily shown by direct calculation that the operators

$$\hat{X}_{\mu\nu}(j) = \psi\hat{\mathbf{a}}^+ X_{\mu\nu}(j)\psi\hat{\mathbf{a}} \tag{3.23}$$

satisfy the commutation relations (3.18) and hence provide the Jordan–Schwinger representation of the group $SO_{n+1}(j)$.

The general element of the algebra $so_{n+1}(j)$ in Jordan–Schwinger representation is written in the form

$$\hat{X}(\mathbf{r},j) = \sum_{\lambda=1}^{n(n+1)/2} r_\lambda \hat{X}_\lambda(j) = \frac{1}{2}(\psi\hat{\mathbf{a}}, \psi\hat{\mathbf{a}}^+)\Delta(\mathbf{r},j)\begin{pmatrix}\psi\hat{\mathbf{a}} \\ \psi\hat{\mathbf{a}}^+\end{pmatrix}$$

$$= \frac{1}{2}(\hat{\mathbf{a}}, \hat{\mathbf{a}}^+)B(\mathbf{r},j)\begin{pmatrix}\hat{\mathbf{a}} \\ \hat{\mathbf{a}}^+\end{pmatrix}, \tag{3.24}$$

where the matrix $\Delta(\mathbf{r},j)$ is given by the equation

$$\Delta(\mathbf{r},j) = \begin{pmatrix} 0 & \epsilon X^T(\mathbf{r},j) \\ X(\mathbf{r},j) & 0 \end{pmatrix}. \tag{3.25}$$

The nonzero matrix elements of $X(\mathbf{r},j)$ are as follows:

$$(X(\mathbf{r},j))_{\nu\mu} = r_\lambda, \quad (X(\mathbf{r},j))_{\mu\nu} = -r_\lambda(\mu,\nu)^2. \tag{3.26}$$

Here λ is connected with μ, ν, $\mu < \nu$ by Eq. (3.20). Using Eq. (3.22) we conclude that the matrix $B(\mathbf{r},j)$ is obtained from the matrix $\Delta(\mathbf{r},j)$ by the following transformation:

$$B(\mathbf{r},j) = \Psi(j)\Delta(\mathbf{r},j)\Psi^{-1}(j). \tag{3.27}$$

We regard first the Cayley–Klein groups $SO_{n+1}(j)$ that are obtained from SO_{n+1} only by the contractions, i.e., when the parameters j_k are equal to the real unit or to the Clifford dual units, $j_k = 1, \iota_k$, $k = 1, 2, \ldots, n$. Then (3.21), (3.22) are identical transformations, $B(\mathbf{r},j) = \Delta(\mathbf{r},j)$, and we have from Eq. (3.5)

$$\Lambda(\mathbf{r},j) = \begin{pmatrix} \Xi(\mathbf{r},j) & 0 \\ 0 & \Xi^T(-\mathbf{r},j) \end{pmatrix}, \tag{3.28}$$

i.e., $\eta = \eta_1 = 0$, $\xi_1 = \Xi^T(-\mathbf{r},j)$, $\xi = \Xi(\mathbf{r},j)$, $\det\xi = \det\Xi(\mathbf{r},j) = 1$, and $\xi^{-1} = \Xi^{-1}(\mathbf{r},j) = \Xi(-\mathbf{r},j)$. Here $\Xi(\mathbf{r},j)$ is the finite rotation matrix of the group $SO_{n+1}(j)$. From Eq. (3.7) the kernel of the finite rotation operator $\hat{U}_g(\mathbf{r},j) = \exp(-\hat{X}(\mathbf{r},j))$ in a coherent state basis is given by

$$U(\alpha^*, \beta, \mathbf{r}, j) = \exp(\alpha^*\Xi(-\mathbf{r},j)\beta). \tag{3.29}$$

In the boson case ($\epsilon = 1$) this kernel is the generating function for the Hermite polynomials $H_{m,n}^{(R(\mathbf{r},j))}(0)$ with the matrix $R(\mathbf{r},j)$ in the form

$$R(\mathbf{r},j) = \begin{pmatrix} 0 & -\Xi(-\mathbf{r},j) \\ -\Xi^T(-\mathbf{r},j) & 0 \end{pmatrix}. \tag{3.30}$$

Contractions of $SO_{n+1}(j)$ under dual values of some of the parameters j_k give rise to limit processes in the generating function (3.29) and hence induce limit processes between Hermite polynomials [Gromov and Man'ko (1991)].

Let the Cayley–Klein groups $SO_{n+1}(j)$ be obtained from SO_{n+1} by both contractions and analytical continuations, i.e., $j_k = 1, \iota_k, i, \ k = 1, 2, \dots, n$. Let us introduce the new parameters $\tilde{j}_k = 1, \iota_k$ as follows: $j_k = i\tilde{j}_k$, if $j_k = i$ and $j_k = \tilde{j}_k$, if $j_k = 1, \iota_k$. The reason for such a redefinition of the parameters is to consider explicitly the analytical continuations and give the associated groups the opportunity of regarding their contractions. The motion integrals matrix $\tilde{\Lambda}(\mathbf{r},j)$ is obtained from the matrix (3.28) as follows [Gromov and Man'ko (1991)]:

$$\tilde{\Lambda}(\mathbf{r},j) = \Psi(j)\Lambda(\mathbf{r},\tilde{\mathbf{j}})\Psi^{-1}(j) = \begin{pmatrix} \xi & \eta \\ \eta_1 & \xi_1 \end{pmatrix}. \tag{3.31}$$

Using Eqs. (3.22), (3.28), and (3.31), we have

$$\begin{aligned} \xi &= \psi_1 \Xi(\mathbf{r},\tilde{j})\psi_1 + \psi_2 \Xi^T(-\mathbf{r},\tilde{j})\psi_2, \\ \xi_1 &= \psi_2 \Xi(\mathbf{r},\tilde{j})\psi_2 + \psi_1 \Xi^T(-\mathbf{r},\tilde{j})\psi_1, \\ \eta &= -\psi_1 \Xi(\mathbf{r},\tilde{j})\psi_2 + \epsilon\psi_2 \Xi^T(-\mathbf{r},\tilde{j})\psi_1, \\ \eta_1 &= \psi_2 \Xi(\mathbf{r},\tilde{j})\psi_1 + \epsilon\psi_1 \Xi^T(-\mathbf{r},\tilde{j})\psi_2, \end{aligned} \tag{3.32}$$

and by Eq. (3.7) obtain the kernel of the finite rotation operator $\hat{U}_g(\mathbf{r},j)$ in a coherent state basis. Note that Eq. (3.7) includes a nonlinear operation of obtaining the inverse matrix ξ^{-1}; therefore, for the kernel we do not have the simple equation as Eqs. (3.27) or (3.31).

To show the effectiveness of the general consideration developed for the Jordan–Schwinger representation of Cayley–Klein groups we shall discuss some groups of low dimensions $SO_2(j_1), SO_3(j), SO_4(j)$ for which it is possible to obtain the explicit form of the finite rotation matrix $\Xi(\mathbf{r},j)$.

3.3.1   *Representations of $SO_2(j_1)$ groups*

The map (3.13), namely $\psi x_0 = x_0, \psi x_1 = j_1 x_1, j_1 = 1, \iota_1, i$, gives the spaces $\mathbf{R}_2(j_1)$ with the metric $\mathbf{x}^2(j_1) = x_0^2 + j_1^2 x_1^2$. Here $\mathbf{R}_2(1)$ is a Euclidean plane, $\mathbf{R}_2(i)$ is the Minkowski (or hyperbolic) plane, and $\mathbf{R}_2(\iota_1)$ is the Galilean plane. Then three groups $SO_2(j_1)$ are as follows: $SO_2(1)$ is the usual rotation group on the plane, $SO_2(i)$ is the group of (one-dimensional) Lorentz transformations, and $SO_2(\iota_1)$ is the group of (one-dimensional) Galilean transformations. Equation (3.17) gives the matrix generator of $SO_2(j_1)$ in the form

$$X_{01}(j_1) = \begin{pmatrix} 0 & -j_1^2 \\ 1 & 0 \end{pmatrix}. \tag{3.33}$$

Then the finite rotation matrix $\Xi(r_1, j_1) = \exp(r_1 X_{01}(j_1))$ is easily obtained:

$$\Xi(r_1, j_1) = \begin{pmatrix} \cos j_1 r_1 & -j_1 \sin j_1 r_1 \\ j_1^{-1} \sin j_1 r_1 & \cos j_1 r_1 \end{pmatrix}. \tag{3.34}$$

A function of dual arguments is defined by its Taylor expansion, therefore $\cos \iota_1 r_1 = 1, \sin \iota_1 r_1 = \iota_1 r_1$, and we have

$$\Xi(r_1, \iota_1) = \begin{pmatrix} 1 & 0 \\ r_1 & 1 \end{pmatrix}, \tag{3.35}$$

i.e., the matrix of Galilean transformation.

When $j_1 = 1$, the operator $\hat{X}_{01}(j_1) = \hat{a}^+ X_{01}(j_1)\hat{a} = \hat{a}_1^+ \hat{a}_0 - j_1^2 \hat{a}_0^+ \hat{a}_1$ provides the Jordan–Schwinger representation of $SO_2(j_1)$ and the kernel of the finite rotation operator $\hat{U}_g(r_1, j_1) = \exp(-r_1 X_{01}(j_1))$ in a coherent state basis is given by, using (3.29) and (3.34),

$$U(\alpha^*, \beta, r_1, j_1) = \exp(\alpha^* \Xi(-r_1, j_1)\beta) = \exp((\alpha_0^* \beta_0 + \alpha_1^* \beta_1) \cos j_1 r_1$$
$$-\alpha_1^* \beta_0 j_1^{-1} \sin j_1 r_1 + \alpha_0^* \beta_1 j_1 \sin j_1 r_1). \tag{3.36}$$

In the boson case, when α_k^*, β_k are the complex variables, the expression (3.36) is the generating function for the Hermite polynomials of four zero-valued variables. We write some first polynomials

$$H_{0,0;0,0}(r_1, j_1) = 1,$$

$$H_{1,1;1,1}(r_1, j_1) = \cos^2 j_1 r_1 - \sin^2 j_1 r_1,$$

$$H_{1,0;1,0}(r_1, j_1) = H_{0,1;0,1}(r_1, j_1) = \cos j_1 r_1, \qquad (3.37)$$

$$H_{0,1;1,0}(r_1, j_1) = -j_1^{-1} \sin j_1 r_1,$$

$$H_{1,0;0,1}(r_1, j_1) = j_1 \sin j_1 r_1.$$

For the Galilean group $SO_2(\iota_1)$ we have $\hat{X}_{01}(\iota_1) = \hat{a}_1^+ \hat{a}_0$ and Eq. (3.36) gives

$$U(\alpha^*, \beta, r_1, \iota_1) = \exp(\alpha_0^* \beta_0 + \alpha_1^* \beta_1 - r_1 \alpha_1^* \beta_0). \qquad (3.38)$$

Then the first Hermite polynomials are

$$H_{0,0;0,0}(r_1, \iota_1) = H_{1,1;1,1}(r_1, \iota_1) = H_{1,0;1,0}(r_1, \iota_1) = H_{0,1;0,1}(r_1, \iota_1) = 1, \qquad (3.39)$$

$$H_{0,1;1,0}(r_1, \iota_1) = -r_1, \quad H_{1,0;0,1}(r_1, \iota_1) = 0.$$

When $j_1 = i$ we introduce the new parameter $\tilde{j}_1$ as $j_1 = i\tilde{j}_1$ and $\tilde{j}_1 = 1, \iota_1$. The case $\tilde{j}_1 = \iota_1$ corresponds to the contraction of the Lorentz group $SO_2(i)$. Equations (3.21) give $\psi \hat{a} = (\hat{a}_0, \hat{a}_1^+), \psi \hat{a}^+ = (\hat{a}_0^+, -\epsilon \hat{a}_1)$ and the operator $\hat{X}_{01}(i\tilde{j}_1) = \psi \mathbf{a}^+ X_{01}(ij_1) \psi \mathbf{a} = j_1^2 \hat{a}_0^+ \hat{a}_1^+ - \epsilon \hat{a}_1 \hat{a}_0$ provides the Jordan–Schwinger representation of $SO_2(i\tilde{j}_1)$. From Eq. (3.22) we obtain the matrices ψ_1 and ψ_2 in the form

$$\psi_1(i) = \begin{pmatrix} 1 & 0 \\ 0 & 0 \end{pmatrix}, \quad \psi_2(i) = \begin{pmatrix} 0 & 0 \\ 0 & -1 \end{pmatrix}. \qquad (3.40)$$

Replacing parameter j_1 in (3.34) by $i\tilde{j}_1$ we have

$$\Xi(r_1, i\tilde{j}_1) = \begin{pmatrix} \cosh \tilde{j}_1 r_1 & \tilde{j}_1 \sinh \tilde{j}_1 r_1 \\ \tilde{j}_1^{-1} \sinh \tilde{j}_1 r_1 & \cosh \tilde{j}_1 r_1 \end{pmatrix}. \qquad (3.41)$$

Then the intermediate matrix $\Lambda(r_1, i\tilde{j}_1)$ is given by Eqs. (3.28) and (3.41). Using it in Eq. (3.31), we obtain the motion integrals matrix $\tilde{\Lambda}(r_1, i\tilde{j}_1)$ of

$SO_2(i\tilde{j}_1)$, namely,

$$\xi = \xi_1 = \begin{pmatrix} 1 & 0 \\ 0 & 1 \end{pmatrix} \cosh \tilde{j}_1 r_1,$$

$$\eta = \begin{pmatrix} 0 & 1 \\ \epsilon & 0 \end{pmatrix} \tilde{j}_1 \sinh \tilde{j}_1 r_1, \qquad (3.42)$$

$$\eta_1 = \begin{pmatrix} 0 & \epsilon \\ 1 & 0 \end{pmatrix} \tilde{j}_1^{-1} \sinh \tilde{j}_1 r_1.$$

The kernel of the operator $\hat{U}_g(r_1, i\tilde{j}_1)$ of the finite Lorentz transformation is given by Eq. (3.7) and is as follows:

$$U(\alpha^*, \beta, r_1, i\tilde{j}_1) = (\cosh \tilde{j}_1 r_1)^{-\epsilon} \exp\{(\cosh \tilde{j}_1 r_1)^{-1}$$
$$\cdot[\alpha_0^* \beta_0 + \alpha_1^* \beta_1 + (\beta_0 \beta_1 - \tilde{j}_1^2 \alpha_0^* \alpha_1^*)\tilde{j}_1^{-1} \sinh \tilde{j}_1 r_1]\}.$$
$$(3.43)$$

Under the contraction ($\tilde{j}_1 = \iota_1$) of the Lorentz group we have for the kernel

$$U(\alpha^*, \beta, r_1, i\iota_1) = \exp(\alpha_0^* \beta_0 + \alpha_1^* \beta_1 + r_1 \beta_0 \beta_1). \qquad (3.44)$$

Comparing the last expression with Eq. (3.38) we conclude that they are different though it follows from Eq. (3.17) that both the group $SO_2(\iota_2)$ and $SO_2(i\iota_1)$ are of the Galilean type. It is the particular case of the general situation [Gromov and Man'ko (1991)]: if a Cayley–Klein group $SO_{n+1}(j)$ is obtained from SO_{n+1} by a k-dimensional contraction, i.e., if k parameters j are equal to the dual units, then there are 2^k different Jordan–Schwinger representations of $SO_{n+1}(j)$.

3.3.2 *Representations of $SO_3(j)$ groups*

The map (3.13), namely $\psi x_0 = x_0, \psi x_1 = j_1 x_1, \psi x_2 = j_1 j_2 x_2, j_1 = 1, \iota_1, i, j_2 = 1, i, \iota_2, i$, gives the nine Cayley–Klein spaces $\mathbf{R}_3(j_1, j_2) = \mathbf{R}_3(j)$ with the metric $\mathbf{x}^2(j) = x_0^2 + j_1^2 x_1^2 + j_1^2 j_2^2 x_2^2$. The nine geometries of the planes of constant curvature are realized [Pimenov (1965)] on the spheres $\mathbf{S}_2(j) = \{\mathbf{x}|\mathbf{x}^2(j) = 1\}$ in the spaces $\mathbf{R}_3(j)$. These geometries according to Fig. 1.2 are as follows: $\mathbf{S}_1(1,1)$ — elliptic; $\mathbf{S}_2(\iota_1, 1)$ — Euclidean; $\mathbf{S}_2(i, 1)$ — Lobachevski (or hyperbolic); $\mathbf{S}_2(1, \iota_2)$ — semielliptic (or co-Euclidean); $\mathbf{S}_2(\iota_2, \iota_2)$ — Galilean; $\mathbf{S}_2(i, \iota_2)$ — semihyperbolic (or co-Minkowski); $\mathbf{S}_2(1, i)$ — anti-de Sitter; $\mathbf{S}_2(\iota_1, i)$ — Minkowski; $\mathbf{S}_2(i, i)$ — de

Sitter. The rotation group $SO_3(j)$ is isomorphic to the motion group of the geometry $\mathbf{S}_2(j)$. We shall call the Cayley–Klein group $SO_3(j)$ by the name of appropriate geometry.

Equation (3.17) gives the matrix generators of $SO_3(j)$ with the help of real matrix elements (however compare with (1.29))

$$X_1(j) = \begin{pmatrix} 0 & -j_1^2 & 0 \\ 1 & 0 & 0 \\ 0 & 0 & 0 \end{pmatrix}, X_2(j) = \begin{pmatrix} 0 & 0 & -j_1^2 j_2^2 \\ 0 & 0 & 0 \\ 1 & 0 & 0 \end{pmatrix},$$

$$X_3(j) = \begin{pmatrix} 0 & 0 & 0 \\ 0 & 0 & -j_2^2 \\ 0 & 1 & 0 \end{pmatrix}. \tag{3.45}$$

The set of generators satisfy the commutation relations

$$[X_1, X_2] = j_1^2 X_3, \quad [X_2, X_3] = j_2^2 X_1, \quad [X_3, X_1] = X_2. \tag{3.46}$$

In accordance with Eq. (3.20) we denote the generators as follows: $X_1 = X_{01}, X_2 = X_{02}, X_3 = X_{12}$. In relation to the general element

$$\mathbf{X}(\mathbf{r}, j) = \sum_{\lambda=1}^{3} r_\lambda X_\lambda(j) = \begin{pmatrix} 0 & -j_1^2 r_1 & -j_1^2 j_2^2 r_2 \\ r_1 & 0 & -j_2^2 r_3 \\ r_2 & r_3 & 0 \end{pmatrix} \tag{3.47}$$

of the algebra $so_3(j)$, the finite rotation of the group $SO_3(j)$ is expressed as

$$\Xi(\mathbf{r}, j) = \exp X(\mathbf{r}, j) = E \cos r + X(\mathbf{r}, j)\frac{\sin r}{r} + X'(\mathbf{r}, j)\frac{1 - \cos r}{r^2}, \tag{3.48}$$

where (compare with (1.39))

$$X'(\mathbf{r}, j) = \begin{pmatrix} j_2^2 r_3^2 & -j_1^2 j_2^2 r_2 r_3 & j_1^2 j_2^2 r_1 r_3 \\ -j_2^2 r_2 r_3 & j_1^2 j_2^2 r_2^2 & -j_1^2 j_2^2 r_1 r_2 \\ r_1 r_3 & -j_1^2 r_1 r_2 & j_1^2 r_1^2 \end{pmatrix}, \tag{3.49}$$

$$\mathbf{r}^2(j) = j_1^2 r_1^2 + j_1^2 j_2^2 r_2^2 + j_2^2 r_3^2.$$

We shall discuss only contractions of the rotation group SO_3, i.e. $j_1 = 1, \iota_1, j_2 = 1, \iota_2$. Then the transformations (3.21), (3.22) are identical and the Jordan–Schwinger representation of the generators (3.45) is given by the operators

$$\hat{X}_1 = \hat{a}_1^+ \hat{a}_0 - j_1^2 \hat{a}_0^+ \hat{a}_1, \quad \hat{X}_2 = \hat{a}_2^+ \hat{a}_0 - j_1^2 j_2^2 \hat{a}_0^+ \hat{a}_2, \quad \hat{X}_3 = \hat{a}_2^+ \hat{a}_1 - j_2^2 \hat{a}_1^+ \hat{a}_2,$$

(3.50)

that satisfy the commutation relations (3.46). The kernel of the finite rotation operator $\hat{U}_g(\mathbf{r}, j) = \exp(-\hat{X}(\mathbf{r}, j))$ in a coherent state basis is given by, using (3.29) and (3.47)–(3.49),

$$\mathbf{U}(\alpha^* \beta, \mathbf{r}, j) = \exp\{\alpha^* \Xi(-\mathbf{r}, j)\beta\}$$

$$= \exp\left\{ \cos r \sum_{k=0}^{2} \alpha_k^* \beta_k - \frac{\sin r}{r}[r_1(\alpha_1^* \beta_0 - j_1^2 \alpha_0^* \beta_1) \right.$$

$$+ r_2(\alpha_2^* \beta_0 - j_1^2 j_2^2 \alpha_0^* \beta_2) + r_3(\alpha_3^* \beta_1 - j_2^2 \alpha_1^* \beta_2)]$$

$$+ \frac{1 - \cos r}{r^2}[j_2^2 r_3^2 \alpha_0^* \beta_0 + j_1^2 j_2^2 r_2^2 \alpha_1^* \beta_1 + j_1^2 r_1^2 \alpha_2^* \beta_2$$

$$- j_2^2 r_2 r_3(\alpha_1^* \beta_0 + j_1^2 \alpha_0^* \beta_1) + r_1 r_3(\alpha_2^* \beta_0 + j_1^2 j_2^2 \alpha_0^* \beta_2)$$

$$\left. - j_1^2 r_1 r_2(\alpha_2^* \beta_1 + j_2^2 \alpha_1^* \beta_2)] \right\}.$$

(3.51)

In the boson case Eq. (3.51) is the generating function for the Hermite polynomials of six zero-valued variables.

3.3.3   *Representations of $SO_4(j)$ groups*

The map (3.13) gives the $3^3 = 27$ Cayley–Klein spaces $\mathbf{R}_4(j)$, $j = (j_1, j_2, j_3)$, $j_k = 1, \iota_k, i$, $k = 1, 2, 3$ with the metric $\mathbf{x}^2(j) = x_0^2 + j_1^2 x_1^2 + j_1^2 j_2^2 x_2^2 + j_1^2 j_2^2 j_3^2 x_3^2$. The three-dimensional spaces of constant curvature are realized on the spheres $\mathbf{S}_3(j) = \{\mathbf{x} \mid \mathbf{x}^2(j) = 1\}$ of the unit real radius in $\mathbf{R}_4(j)$. Some of these spaces are well known, for example, $\mathbf{S}_3(i, 1, 1)$ — Lobachevski; $\mathbf{S}_3(\iota_1, 1, 1)$ — Euclidean, $\mathbf{S}_3(\iota_1, i, 1)$ — Minkowski; $\mathbf{S}_3(\iota_1, \iota_2, 1)$ — Galilean, and some do not have special names.

The six matrix generators of $SO_4(j)$ are given by Eq. (3.17) as follows:

$$X_1 = X_{01} = \begin{pmatrix} 0 & -j_1^2 & 0 & 0 \\ 1 & 0 & 0 & 0 \\ 0 & 0 & 0 & 0 \\ 0 & 0 & 0 & 0 \end{pmatrix}, \quad X_2 = X_{02} = \begin{pmatrix} 0 & 0 & -j_1^2 j_2^2 & 0 \\ 0 & 0 & 0 & 0 \\ 1 & 0 & 0 & 0 \\ 0 & 0 & 0 & 0 \end{pmatrix},$$

$$X_3 = X_{03} = \begin{pmatrix} 0 & 0 & 0 & -j_1^2 j_2^2 j_3^2 \\ 0 & 0 & 0 & 0 \\ 0 & 0 & 0 & 0 \\ 1 & 0 & 0 & 0 \end{pmatrix}, \quad X_4 = X_{12} = \begin{pmatrix} 0 & 0 & 0 & 0 \\ 0 & 0 & -j_2^2 & 0 \\ 0 & 1 & 0 & 0 \\ 0 & 0 & 0 & 0 \end{pmatrix},$$

$$X_5 = X_{13} = \begin{pmatrix} 0 & 0 & 0 & 0 \\ 0 & 0 & 0 & -j_2^2 j_3^2 \\ 0 & 0 & 0 & 0 \\ 0 & 1 & 0 & 0 \end{pmatrix}, \quad X_6 = X_{23} = \begin{pmatrix} 0 & 0 & 0 & 0 \\ 0 & 0 & 0 & 0 \\ 0 & 0 & 0 & -j_3^2 \\ 0 & 0 & 1 & 0 \end{pmatrix},$$

$$(3.52)$$

and due to Eq. (3.18) satisfy the commutation relations

$$[X_1, X_2] = j_1^2 X_4, \quad [X_1, X_3] = j_1^2 X_5, \quad [X_2, X_3] = j_1^2 j_2^2 X_6,$$

$$[X_1, X_4] = -X_2, \quad [X_1, X_5] = -X_3, \quad [X_2, X_6] = -X_3,$$

$$[X_2, X_4] = j_2^2 X_1, \quad [X_3, X_5] = j_2^2 j_3^2 X_1, \quad [X_3, X_6] = j_3^2 X_2, \quad (3.53)$$

$$[X_4, X_5] = j_2^2 X_6, \quad [X_4, X_6] = -X_5, \quad [X_5, X_6] = j_3^2 X_4.$$

Let us introduce the new denominations for the group parameters, namely, $r_4 = -s_3, r_5 = s_2, r_6 = -s_1$. Then the general element of the algebra $so_4(j)$ is

$$X(\mathbf{r}, \mathbf{s}, j) = \sum_{k=1}^{3} r_k X_k(j) - s_3 X_4(j) + s_2 X_5(j) - s_1 X_6(j)$$

$$= \begin{pmatrix} 0 & -j_1^2 r_1 & -j_1^2 j_2^2 r_2 & -j_1^2 j_2^2 j_3^2 r_3 \\ r_1 & 0 & j_2^2 s_3 & -j_2^2 j_3^2 s_2 \\ r_2 & -s_3 & 0 & j_3^2 s_1 \\ r_3 & s_2 & -s_1 & 0 \end{pmatrix} \quad (3.54)$$

and the finite rotation matrix of the group $SO_4(j)$ is given by the following equation:

$$\Xi(\mathbf{r}, \mathbf{s}, j) = \frac{EA + XB + (\mathbf{r}, \mathbf{s})X_1 C + X^2 D}{\sqrt{(r^2 + s^2)^2 - 4j_1^2 j_2^2 (\mathbf{r}, \mathbf{s})^2}}, \quad (3.55)$$

where E is a four-dimensional unit matrix, the matrix X is given by Eq. (3.54), the matrices X_1 and X_2 are in the form

$$X_1 = \begin{pmatrix} 0 & -j_1^2 j_3^2 s_1 & -j_1^2 j_2^2 j_3^2 s_2 & -j_1^2 j_2^2 j_3^2 s_3 \\ j_3^2 s_1 & 0 & j_1^2 j_2^2 j_3^2 r_3 & -j_1^2 j_2^2 j_3^2 r_2 \\ j_3^2 s_2 & -j_1^2 j_3^2 r_3 & 0 & j_1^2 j_3^2 r_1 \\ s_3 & j_1^2 r_2 & -j_1^2 r_1 & 0 \end{pmatrix}, \tag{3.56}$$

$$X^2 = \begin{pmatrix} -r^2 & j_1^2 j_2^2 (\mathbf{r} \times \mathbf{s})_1 & j_1^2 j_2^2 (\mathbf{r} \times \mathbf{s})_2 \\ j_2^2 (\mathbf{r} \times \mathbf{s})_1 & j_3^2 s_1^2 - j_1^2 r_1^2 - s^2 & j_2^2 j_3^2 s_1 s_2 - j_1^2 j_2^2 r_1 r_2 \\ (\mathbf{r} \times \mathbf{s})_2 & j_3^2 s_1 s_2 - j_1^2 r_1 r_2 & j_2^2 j_3^2 s_2^2 - j_1^2 j_2^2 r_2^2 - s^2 \\ \mathbf{r} \times \mathbf{s})_3 & s_1 s_3 - j_1^2 r_1 r_3 & j_2^2 s_2 s_3 - j_1^2 j_2^2 r_2 r_3 \end{pmatrix}$$

$$\begin{pmatrix} j_1^2 j_2^2 j_3^2 (\mathbf{r} \times \mathbf{s})_3 \\ j_2^2 j_3^2 s_1 s_3 - j_1^2 j_2^2 j_3^2 r_1 r_3 \\ j_2^2 j_3^2 s_2 s_3 - j_1^2 j_2^2 j_3^2 r_2 r_3 \\ j_2^2 s_3^2 - j_1^2 j_2^2 j_3^2 r_3^2 - s^2 \end{pmatrix}. \tag{3.57}$$

The functions A, B, C, D in Eq. (3.55) are equal to

$$A = z_1 \cos \sqrt{-z_2} - z_2 \cos \sqrt{-z_1},$$

$$B = z_1 \frac{\sin \sqrt{-z_1}}{\sqrt{-z_1}} - z_2 \frac{\sin \sqrt{-z_2}}{\sqrt{-z_2}},$$

$$C = \frac{\sin \sqrt{-z_1}}{\sqrt{-z_1}} - \frac{\sin \sqrt{-z_2}}{\sqrt{-z_2}}, \tag{3.58}$$

$$D = \cos \sqrt{-z_1} - \cos \sqrt{-z_2},$$

where

$$z_{1,2} = -\frac{1}{2} \left(r^2 + s^2 \mp \sqrt{(r^2 + s^2)^2 - 4 j_1^2 j_3^2 (\mathbf{r}, \mathbf{s})} \right). \tag{3.59}$$

We use the following denominations:

$$r^2 = j_1^2 r_1^2 + j_1^2 j_2^2 r_2^2 + j_1^2 j_2^2 j_3^2 r_3^2, \quad s^2 = j_3^2 s_1^2 + j_2^2 j_3^2 s_2^2 + j_2^2 s_3^2,$$

$$(\mathbf{r}, \mathbf{s}) = r_1 s_1 + j_2^2 (r_2 s_2 + r_3 s_3), \quad (\mathbf{r} \times \mathbf{s})_1 = r_2 s_3 - j_3^2 r_3 s_2, \tag{3.60}$$

$$(\mathbf{r} \times \mathbf{s})_2 = j_3^2 r_3 s_1 - r_1 s_3, \quad (\mathbf{r} \times \mathbf{s})_3 = r_1 s_2 - r_2 s_1.$$

In the case of only contractions of the orthogonal group SO_4, i.e., $j_k = 1, \iota_k$, $k = 1, 2, 3$, the Jordan–Schwinger representation of the generators (3.52) is given by the expression $\hat{X}_k(j) = \hat{a}^+ X_k(j)\hat{a}$ and the kernel of the finite rotation operator is obtained by Eq. (3.29) with help of the matrix (3.55). We shall not write out this kernel.

We point out the connection of the Jordan–Schwinger representations with the properties of quantum systems. The general element $\hat{X}(\mathbf{r}, j)$ of the algebra $so_{n+1}(j)$ in the Jordan–Schwinger representation is the linear function of the second quantized generators of $SO_{n+1}(j)$, therefore the replacement of the group parameters $\mathbf{r}$ by $(i/\hbar)t\mathbf{r}$, where t is the time variable, transform the finite rotation operator $\hat{U}_g = \exp(-\hat{X}(\mathbf{r}, j))$ into the evolution operator $\hat{U} = \exp(-(i/\hbar)t H) = \exp(-(i/\hbar)t\hat{X}(\mathbf{r}, j))$ of the quantum system with the Hamiltonian $\hat{H} = \hat{X}(\mathbf{r}, j)$. The last quantum systems are called the group quantum systems [Gromov and Man'ko (1991)]. In the case of stationary systems (when the group parameters $\mathbf{r}$ do not depend on the time t) the kernel of the finite rotation operator is transformed into the matrix elements of the evolution operator (or Green's function) of corresponding quantum system as follows: $G(\alpha^*, \beta, \mathbf{r}, j, t) = U(\alpha^*, \beta, (i/\hbar)t\mathbf{r}, j)$. Thus, by investigating the Jordan–Schwinger representations of set of groups $SO_{n+1}(j)$, we simultaneously investigate the set of stationary quantum systems, corresponding to $SO_{n+1}(j)$.

3.4 The Jordan–Schwinger representations of the special unitary Cayley–Klein groups

Let us use the map (1.51) of the $(n+1)$-dimensional complex space $\mathbf{C}_{n+1}$ into the complex Cayley–Klein space $\mathbf{C}_{n+1}(j)$ as follows:

$$
\begin{aligned}
\psi : \quad & \mathbf{C}_{n+1} \to \mathbf{C}_{n+1}(j), \\
& \psi z_0^* = z_0, \quad \psi z_k^* = (0, k) z_k,
\end{aligned}
\tag{3.61}
$$

where $k = 1, 2, \ldots, n$; z_0, z_k are the complex Cartesian coordinates, $j = (j_1, j_2, \ldots, j_n)$ and each of the parameters j_k may be equal to the real unit 1, the Clifford dual unit ι_k, or the imaginary unit i. The quadratic form $(\mathbf{z}^*, \mathbf{z}^*) = \sum_{m=0}^{n} |z_m^*|^2$ of $\mathbf{C}_{n+1}$ transforms under the map (3.61) into the quadratic form (1.52) of the complex Cayley–Klein space $\mathbf{C}_{n+1}(j)$.

The unitary Cayley–Klein group $SU_{n+1}(j)$ consists of all the transformations of the space $\mathbf{C}_{n+1}(j)$, which keep the quadratic form (1.52) invariant. The map (3.61) induces the transformation of the special unitary group SU_{n+1} into the group $SU_{n+1}(j)$. All $(n+1)^2 - 1$ generators of SU_{n+1} are Hermite matrices. The commutation relations for these Hermite generators are very complicated and usually the matrix generators $\tilde{Y}_{km}$, $k, m = 0, 1, \ldots, n$ of the general linear group $GL_{n+1}(\mathbb{R})$ are used. The only nonzero matrix element of $\tilde{Y}_{km}$ is $(\tilde{Y}_{km})_{km} = 1$. The generators $\tilde{Y}$ of $GL_{n+1}(\mathbb{R})$ satisfy the commutation relations

$$[\tilde{Y}_{km}, \tilde{Y}_{pq}] = \delta_{mp}\tilde{Y}_{kq} - \delta_{kq}\tilde{Y}_{mp}. \tag{3.62}$$

The independent Hermite generators of SU_{n+1} are defined by the equations

$$\tilde{Q}_{\mu\nu} = \frac{i}{2}(\tilde{Y}_{\mu\nu} + \tilde{Y}_{\nu\mu}), \quad \tilde{L}_{\mu\nu} = \frac{1}{2}(\tilde{Y}_{\nu\mu} - \tilde{Y}_{\mu\nu}),$$
$$\tilde{P}_k = \frac{i}{2}(\tilde{Y}_{k-1,k-1} - \tilde{Y}_{kk}), \tag{3.63}$$

where $\mu = 0, 1, \ldots, n-1$, $\nu = \mu+1, \mu+2, \ldots, n$. The matrix generators $\tilde{Y}$ are transformed under the map (3.61) as follows:

$$\tilde{Y}_{\mu\nu}(j) = (\mu, \nu)\tilde{Y}_{\mu\nu}(\rightarrow) = (\mu, \nu)^2\tilde{Y}_{\mu\nu},$$
$$\tilde{Y}_{\nu\mu}(j) = (\mu, \nu)\tilde{Y}_{\nu\mu}(\rightarrow) = \tilde{Y}_{\nu\mu}, \quad \mu < \nu, \tag{3.64}$$
$$Y_{kk}(j) = \tilde{Y}_{kk},$$

where $\tilde{Y}_{\mu\nu}(\rightarrow)$ and $\tilde{Y}_{\nu\mu}(\rightarrow)$ denote the transformed generators $\tilde{Y}_{\mu\nu}, \tilde{Y}_{\nu\mu}$ with the following nonzero matrix elements:

$$(\tilde{Y}_{\mu\nu}(\rightarrow))_{\mu\nu} = (\mu, \nu)(\tilde{Y}_{\mu\nu})_{\mu\nu} = (\mu, \nu),$$
$$(\tilde{Y}_{\nu\mu}(\rightarrow))_{\nu\mu} = (\mu, \nu)^{-1}(\tilde{Y}_{\nu\mu})_{\nu\mu} = (\mu, \nu)^{-1}. \tag{3.65}$$

The generators (3.64) satisfy the commutation relations

$$[Y_{km}, Y_{pq}] = (l_1, l_2)(l_3, l_4)\left(\delta_{mp}Y_{kq}(l_5, l_6)^{-1} - \delta_{kq}Y_{mp}(l_7, l_8)^{-1}\right), \tag{3.66}$$

where $l_1 = 1 + \min(k, m)$, $l_2 = \max(k, m)$, $l_3 = 1 + \min(p, q)$, $l_4 = \max(p, q)$, $l_5 = 1 + \min(k, q)$, $l_6 = \max(k, q)$, $l_7 = 1 + \min(m, p)$, $l_8 = \max(m, p)$. The same laws of transformations as in Eq. (3.64) are held for the Hermite generators (3.63). Then we obtain the matrix generators of the unitary Cayley–Klein group $SU_{n+1}(j)$ in the form

$$Q_{\mu\nu}(j) = (\mu, \nu)\tilde{Q}_{\mu\nu}(\rightarrow) = \frac{i}{2}(Y_{\nu\mu}(j) + Y_{\mu\nu}(j)) = \frac{i}{2}\left(\tilde{Y}_{\nu\mu} + \tilde{Y}_{\mu\nu}(\mu, \nu)^2\right),$$

$$L_{\mu\nu}(j) = (\mu, \nu)\tilde{L}_{\mu\nu}(\rightarrow) = \frac{1}{2}(Y_{\nu\mu}(j) - Y_{\mu\nu}(j)) = \frac{1}{2}\left(\tilde{Y}_{\nu\mu} - \tilde{Y}_{\mu\nu}(\mu, \nu)^2\right),$$

$$P_k(j) = \tilde{P}_k = \frac{i}{2}\left(\tilde{Y}_{k-1,k-1} - \tilde{Y}_{kk}\right), \quad k = 1, 2, \ldots, n. \tag{3.67}$$

The commutation relations of these generators may be derived with the help of Eq. (3.66), but they are very cumbersome and we do not write them here.

The finite group transformation

$$\Xi(\mathbf{r}, \mathbf{s}, \mathbf{w}, j) = \exp Z(\mathbf{r}, \mathbf{s}, \mathbf{w}, j) \tag{3.68}$$

of $SU_{n+1}(j)$ corresponds to the general element

$$Z(\mathbf{r}, \mathbf{s}, \mathbf{w}, j) = \sum_{\lambda=1}^{n(n+1)/2} (r_\lambda Q_\lambda(j) + s_\lambda L_\lambda(j)) + \sum_{k=1}^{n} \omega_k P_k(j) \tag{3.69}$$

of the algebra $su_{n+1}(j)$, where $r_\lambda, s_\lambda, \omega_k$ are real group parameters and index λ is connected with the indices μ, ν by Eq. (3.20). The explicit form of the finite transformation $\Xi(\mathbf{r}, \mathbf{s}, \mathbf{w}, j)$ may be obtained for the groups $SU_2(j_1)$ and $SU_3(j_1, j_2)$.

The Jordan–Schwinger representations of $SU_{n+1}(j)$ are provided by the operators

$$\hat{Q}_{\mu\nu}(j) = \psi\hat{\mathbf{a}}^+ Q_{\mu\nu}(j)\psi\hat{\mathbf{a}}, \quad \hat{L}_{\mu\nu}(j) = \psi\hat{\mathbf{a}}^+ L_{\mu\nu}(j)\psi\hat{\mathbf{a}},$$
$$P_k(j) = \psi\hat{\mathbf{a}}^+ P_k(j)\psi\hat{\mathbf{a}}, \quad \mu < \nu, \ k = 1, 2, \ldots, n, \tag{3.70}$$

where the transformed sets $\psi\hat{\mathbf{a}}^+, \psi\hat{\mathbf{a}}$ of the creation and annihilation operators are given by Eqs. (3.21) and (3.22). Indeed, it is verified by direct calculations that the operators $\hat{Y}(j) = \psi\hat{\mathbf{a}}^+ Y(j)\psi\hat{\mathbf{a}}$ satisfy the commutation relations (3.66) and we may conclude that the operators (3.70) satisfy the commutation relations of the group $SU_{n+1}(j)$. The finite group transformation is represented by the operator

$$\hat{U}_g(\mathbf{r}, \mathbf{s}, \mathbf{w}, j) = \exp(-\hat{Z}(\mathbf{r}, \mathbf{s}, \mathbf{w}, j)), \tag{3.71}$$

where the operator $\hat{Z}$ is given by Eq. (3.69) with the generators $Q_\lambda(j), L_\lambda(j), P_k(j)$ replaced by the operators $\hat{Q}_\lambda(j), \hat{L}_\lambda(j)$ and $\hat{P}_k(j)$, respectively. The kernel of the operator $\hat{U}_g(\mathbf{r}, \mathbf{s}, \mathbf{w}, j)$ in a coherent state basis is obtained quite analogously to the case of orthogonal groups. When the Cayley–Klein groups $SU_{n+1}(j)$ are obtained from SU_{n+1} by only contractions ($j_k = 1, \iota_k, \ k = 1, 2, \ldots, n$) this kernel is given by

$$U(\alpha^*, \beta, \mathbf{r}, \mathbf{s}, \mathbf{w}, j) = \exp(\alpha^* \Xi(-\mathbf{r}, -\mathbf{s}, -\mathbf{w}, j)\beta), \tag{3.72}$$

(compare with Eq. (3.29)). When the groups $SU_{n+1}(j)$ are obtained from SU_{n+1} by both contractions and analytical continuations the kernel is given by the following equation:

$$U(\alpha^*, \beta, \mathbf{r}, \mathbf{s}, \mathbf{w}, j) = (\det \xi)^{-\epsilon/2} \exp$$

$$\left(\frac{1}{2} \alpha^* \xi^{-1} \eta \alpha^* + \alpha^* \xi^{-1} \beta + \frac{1}{2} \epsilon \beta \eta_1 \xi^{-1} \beta \right), \quad (3.73)$$

where the matrices ξ, η, η_1 are expressed through the matrix (3.68) by Eqs. (3.32).

We shall discuss in detail two sets of unitary Cayley–Klein groups $SU_2(j_1)$ and $SU_3(j_1, j_2)$ for which it is possible to obtain the explicit form of the finite group transformation matrix $\Xi(\mathbf{r}, \mathbf{s}, \mathbf{w}, j)$.

3.4.1   *Representations of $SU_2(j_1)$ groups*

The map (3.61), namely $\psi z_0^* = z_0$, $\psi z_1^* = j_1 z_1$, $j_1 = 1, \iota_1, i$, gives the complex space $\mathbf{C}_2(j_1)$, with the quadratic form $(\mathbf{z}, \mathbf{z}) = |z_0|^2 + j_1^2 |z_1|^2$. The transformations belonging to the group $SU_2(j_1)$ keep this quadratic form invariant. The matrices (3.64) are as follows:

$$Y_{00} = \begin{pmatrix} 1 & 0 \\ 0 & 0 \end{pmatrix}, \ Y_{11} = \begin{pmatrix} 0 & 0 \\ 0 & 1 \end{pmatrix}, \ Y_{01} = \begin{pmatrix} 0 & j_1^2 \\ 0 & 0 \end{pmatrix}, \ Y_{10} = \begin{pmatrix} 0 & 0 \\ 1 & 0 \end{pmatrix}, \quad (3.74)$$

and the commutation relations

$$[Y_{00}, Y_{11}] = 0, \quad [Y_{00}, Y_{01}] = Y_{01}, \quad [Y_{00}, Y_{10}] = -Y_{10},$$
$$[Y_{11}, Y_{01}] = -Y_{01}, \quad [Y_{11}, Y_{10}] = Y_{10}, \quad [Y_{01}, Y_{10}] = j_1^2 Y_{00} \quad (3.75)$$

are satisfied. The matrix generators of $SU_2(j_1)$ are given by Eq. (3.67) in the form

$$P_1 = \frac{i}{2} \begin{pmatrix} 1 & 0 \\ 0 & -1 \end{pmatrix}, \quad Q_{01} = \frac{i}{2} \begin{pmatrix} 0 & j_1^2 \\ 1 & 0 \end{pmatrix}, \quad L_{01} = \frac{1}{2} \begin{pmatrix} 0 & -j_1^2 \\ 1 & 0 \end{pmatrix}, \quad (3.76)$$

and satisfy the following commutation relations

$$[P_1, Q_{01}] = L_{01}, \quad [L_{01}, P_1] = Q_{01}, \quad [Q_{01}, L_{01}] = j_1^2 P_1. \quad (3.77)$$

The generators (3.76) for $j_1 = 1$ are equal up to a coefficient to the Pauli matrices. The general element of algebra $su_2(j_1)$ in accordance with

Eq. (3.69) is given by

$$\mathbf{Z}(r_1, s_1, w_1, j_1) = r_1 Q_{01} + s_1 L_{01} + w_1 P_1$$

$$= \frac{1}{2} \begin{pmatrix} iw_1 & -j_1^2(s_1 - ir_1) \\ s_1 + ir_1 & -iw_1 \end{pmatrix}, \tag{3.78}$$

and for the finite group transformation matrix we have

$$\Xi(r_1, s_1, w_1, j_1) = E_2 \cos \frac{v}{2} + Z \frac{2}{v} \sin \frac{v}{2}$$

$$= \begin{pmatrix} \cos \frac{v}{2} + i\frac{w_1}{v} \sin \frac{v}{2} & -j_1^2(s_1 - ir_1)\frac{1}{v} \sin \frac{v}{2} \\ (s_1 + ir_1)\frac{1}{v} \sin \frac{v}{2} & \cos \frac{v}{2} - i\frac{w_1}{v} \sin \frac{v}{2} \end{pmatrix}, \tag{3.79}$$

where

$$v^2(j_1) = w_1^2 + j_1^2(r_1^2 + s_1^2). \tag{3.80}$$

When $j_1 = 1, \iota_1$ the following operators:

$$\hat{Q}_{01}(j_1) = \hat{a}^+ Q_{01}(j_1)\hat{a} = \tfrac{i}{2}(\hat{a}_1^+ \hat{a}_0 + j_1^2 \hat{a}_0^+ \hat{a}_1),$$
$$\hat{L}_{01}(j_1) = \hat{a}^+ L_{01}(j_1)\hat{a} = \tfrac{1}{2}(\hat{a}_1^+ \hat{a}_0 - j_1^2 \hat{a}_0^+ \hat{a}_1), \tag{3.81}$$
$$\hat{P}_1(j_1) = \hat{a}^+ P_1(j_1)\hat{a} = \tfrac{i}{2}(\hat{a}_0^+ \hat{a}_0 - \hat{a}_1^+ \hat{a}_1),$$

satisfy the commutation relations (3.77) and therefore provide the Jordan–Schwinger representation of $SU_2(j_1)$. The kernel of the finite group transformation operator is obtained by Eq. (3.72) with help of the matrix (3.79) and is given in the form

$$U(\alpha^*, \beta, j_1) = \exp\left\{ \alpha_0^* \beta_0 \left(\cos \frac{v}{2} - i\frac{w_1}{v} \sin \frac{v}{2} \right) + \alpha_1^* \beta_1 \left(\cos \frac{v}{2} + i\frac{w_1}{v} \sin \frac{v}{2} \right) \right.$$

$$\left. -\alpha_1^* \beta_0(s_1 + ir_1)\frac{1}{v} \sin \frac{v}{2} + j_1^2 \alpha_0^* \beta_1(s_1 - ir_1)\frac{1}{v} \sin \frac{v}{2} \right\}. \tag{3.82}$$

For the contracted group $SU_2(\iota_1)$ we have from Eq. (3.80) $v = w_1$ and from Eq. (3.82)

$$U(\alpha^*, \beta, \iota_1) = \exp\left\{ \alpha_0^* \beta_0 e^{-(i/2)w_1} + \alpha_1^* \beta_1 e^{(i/2)w_1} \right.$$

$$\left. -\alpha_1^* \beta_0(s_1 + ir_1)\frac{1}{w_1} \sin \frac{w_1}{2} \right\}. \tag{3.83}$$

When $j_1 = i$ we introduce the new parameter $\tilde{j}_1$ as $j_1 = i\tilde{j}_1, \tilde{j} = i, \iota_1$. The case $j_1 = \iota_1$ corresponds to the contraction of the pseudounitary group

$SU_2(i) \equiv SU(1,1)$. Equations (3.21) give $\psi\hat{\mathbf{a}} = (\hat{a}_0, \hat{a}_1^+), \psi\hat{\mathbf{a}} = (\hat{a}_0^+, -\epsilon\hat{a}_1)$. Then from Eq. (3.70) we obtain the operators

$$\hat{Q}_{01}(\tilde{i}\tilde{j}_1) = \psi\hat{\mathbf{a}}^+ Q_{01}(\tilde{i}\tilde{j}_1)\psi\hat{\mathbf{a}} = -\frac{i}{2}(\epsilon\hat{a}_1\hat{a}_0 + j_1^2\hat{a}_0^+\hat{a}_1^+),$$

$$\hat{L}_{01}(\tilde{i}\tilde{j}_1) = \psi\hat{\mathbf{a}}^+ L_{01}(\tilde{i}\tilde{j}_1)\psi\hat{\mathbf{a}} = -\frac{1}{2}(\epsilon\hat{a}_1\hat{a}_0 - j_1^2\hat{a}_0^+\hat{a}_1^+), \qquad (3.84)$$

$$\hat{P}_1(\tilde{i}\tilde{j}_1) = \psi\hat{\mathbf{a}}^+ P_1(\tilde{i}\tilde{j}_1)\psi\hat{\mathbf{a}} = \frac{i}{2}(\hat{a}_0^+\hat{a}_0 + \epsilon\hat{a}_1\hat{a}_1^+),$$

that provide the Jordan–Schwinger representation of $SU_2(\tilde{i}\tilde{j}_1)$. The diagonal matrices $\psi_1(i), \psi_2(i)$ in Eq. (3.22) are of the form

$$\psi_1(i) = \begin{pmatrix} 1 & 0 \\ 0 & 0 \end{pmatrix}, \quad \psi_2(i) = \begin{pmatrix} 0 & 0 \\ 0 & -1 \end{pmatrix}, \qquad (3.85)$$

and using Eqs. (3.32) and (3.79) we obtain the matrices ξ, η, η_1

$$\xi = \begin{pmatrix} 1 & 0 \\ 0 & 1 \end{pmatrix} \left(\cos\frac{\tilde{v}}{2} + i\frac{w_1}{\tilde{v}}\sin\frac{\tilde{v}}{2} \right),$$

$$\eta = \begin{pmatrix} 0 & 1 \\ 1 & 0 \end{pmatrix} \tilde{j}_1^2(s_1 - ir_1)\frac{1}{\tilde{v}}\sin\frac{\tilde{v}}{2},$$

$$\eta_1 = \begin{pmatrix} 0 & 1 \\ 1 & 0 \end{pmatrix} (s_1 + ir_1)\frac{1}{\tilde{v}}\sin\frac{\tilde{v}}{2}, \qquad (3.86)$$

$$\tilde{v}^2 = w_1^2 - j_1^2(r_1^2 + s_1^2).$$

After some calculations we find the matrices $\xi^{-1}, \xi^{-1}\eta, \eta_1\xi^{-1}$ and using Eq. (3.73) obtain the kernel of the finite group transformation operator of $SU_2(\tilde{i}\tilde{j}_1)$ in the form

$$U(\alpha^*, \beta, \tilde{i}\tilde{j}_1) = \left(\cos\frac{\tilde{v}}{2} + i\frac{w_1}{\tilde{v}}\sin\frac{\tilde{v}}{2} \right)^{-\epsilon} \exp\left\{ \left(\cos\frac{\tilde{v}}{2} + i\frac{w_1}{\tilde{v}}\sin\frac{\tilde{v}}{2} \right)^{-1} \right.$$

$$\times \left[\alpha_0^*\beta_0 + \alpha_1^*\beta_1 - \frac{1}{2}(\alpha_0^*\alpha_1^* + \alpha_1^*\alpha_0^*)\tilde{j}_1^2(s_1 - ir_1)\frac{1}{\tilde{v}}\sin\frac{\tilde{v}}{2} \right.$$

$$\left. \left. + \frac{1}{2}\epsilon(\beta_0\beta_1 + \beta_1\beta_0)(s_1 + ir_1)\frac{1}{\tilde{v}}\sin\frac{\tilde{v}}{2} \right] \right\}. \qquad (3.87)$$

For the contracted group $SU_2(i\iota_1)$ we have from Eq. (3.86) $v = w_1$ and from Eq. (3.87)

$$U(\alpha^*, \beta, i\iota_1) = e^{-(i/2)\epsilon w_1} \exp\left\{ e^{-(i/2)w_1} \left[\alpha_0^*\beta_0 + \alpha_1^*\beta_1 + \frac{1}{2}\epsilon(\beta_0\beta_1 + \beta_1\beta_0) \right. \right.$$

$$\left. \left. \times (s_1 + ir_1)\frac{1}{w_1}\sin\frac{w_1}{2} \right] \right\}. \qquad (3.88)$$

Comparing Eq. (3.88) with Eq. (3.83) they provide different Jordan–Schwinger representations of the same group $SU_2(\iota_1) = SU_2(i\iota_1)$. Notice that some works [Dattoli, Richetta and Torre (1988)], [de Prunele (1988)] were devoted to the connections between $SU(1,1)$ and $SU(2)$, and to the evolution of $SU(2)$ coherent state from a unified point of view.

3.4.2 *Representations of $SU_3(j_1, j_2)$ groups*

The $SU_3(j_1, j_2)$ group consists of all transformations of $\mathbf{C}_{2+1}(j_1, j_2)$ keeping invariant the quadratic form $(\mathbf{z}, \mathbf{z}) = |z_0|^2 + j_1^2|z_1|^2 + j_1^2 j_2^2|z_2|^2$. The matrix generators of the general linear group $GL_3(j_1, j_2; \mathbb{R})$ are given by Eq. (3.64) in the form

$$
Y_{00} = \begin{pmatrix} 1 & 0 & 0 \\ 0 & 0 & 0 \\ 0 & 0 & 0 \end{pmatrix}, \quad
Y_{11} = \begin{pmatrix} 0 & 0 & 0 \\ 0 & 1 & 0 \\ 0 & 0 & 0 \end{pmatrix}, \quad
Y_{22} = \begin{pmatrix} 0 & 0 & 0 \\ 0 & 0 & 0 \\ 0 & 0 & 1 \end{pmatrix},
$$

$$
Y_{10} = \begin{pmatrix} 0 & 0 & 0 \\ 1 & 0 & 0 \\ 0 & 0 & 0 \end{pmatrix}, \quad
Y_{01} = \begin{pmatrix} 0 & j_1^2 & 0 \\ 0 & 0 & 0 \\ 0 & 0 & 0 \end{pmatrix}, \quad
Y_{20} = \begin{pmatrix} 0 & 0 & 0 \\ 0 & 0 & 0 \\ 1 & 0 & 0 \end{pmatrix}, \quad (3.89)
$$

$$
Y_{02} = \begin{pmatrix} 0 & 0 & j_1^2 j_2^2 \\ 0 & 0 & 0 \\ 0 & 0 & 0 \end{pmatrix}, \quad
Y_{21} = \begin{pmatrix} 0 & 0 & 0 \\ 0 & 0 & 0 \\ 0 & 1 & 0 \end{pmatrix}, \quad
Y_{12} = \begin{pmatrix} 0 & 0 & 0 \\ 0 & 0 & j_2^2 \\ 0 & 0 & 0 \end{pmatrix}.
$$

Then the matrix generators of $SU_3(j_1, j_2)$ are given by Eq. (3.67) as follows:

$$
P_1 = \frac{i}{2} \begin{pmatrix} 1 & 0 & 0 \\ 0 & -1 & 0 \\ 0 & 0 & 0 \end{pmatrix}, \quad
P_2 = \frac{i}{2} \begin{pmatrix} 0 & 0 & 0 \\ 0 & 1 & 0 \\ 0 & 0 & -1 \end{pmatrix},
$$

$$
Q_1 = Q_{01} = \frac{i}{2} \begin{pmatrix} 0 & j_1^2 & 0 \\ 1 & 0 & 0 \\ 0 & 0 & 0 \end{pmatrix}, \quad
L_1 = L_{01} = \frac{1}{2} \begin{pmatrix} 0 & -j_1^2 & 0 \\ 1 & 0 & 0 \\ 0 & 0 & 0 \end{pmatrix},
$$

$$
Q_2 = Q_{02} = \frac{i}{2} \begin{pmatrix} 0 & 0 & j_1^2 j_2^2 \\ 0 & 0 & 0 \\ 1 & 0 & 0 \end{pmatrix}, \quad
L_2 = L_{02} = \frac{1}{2} \begin{pmatrix} 0 & 0 & -j_1^2 j_2^2 \\ 0 & 0 & 0 \\ 1 & 0 & 0 \end{pmatrix},
$$

$$Q_3 = Q_{12} = \frac{i}{2} \begin{pmatrix} 0 & 0 & 0 \\ 0 & 0 & j_2^2 \\ 0 & 1 & 0 \end{pmatrix}, \quad L_3 = L_{12} = \frac{1}{2} \begin{pmatrix} 0 & 0 & 0 \\ 0 & 0 & -j_2^2 \\ 0 & 1 & 0 \end{pmatrix}. \quad (3.90)$$

They satisfy the commutation relations

$$[P_1, P_2] = 0, \quad [P_1, Q_1] = L_1, \quad [P_1, L_1] = -Q_1,$$

$$[P_1, Q_2] = \frac{1}{2} L_2, \quad [P_1, L_2] = -\frac{1}{2} Q_2, \quad [P_1, Q_3] = -\frac{1}{2} L_3,$$

$$[P_1, L_3] = \frac{1}{2} Q_3, \quad [P_2, Q_1] = -\frac{1}{2} L_1, \quad [P_2, L_1] = \frac{1}{2} Q_1,$$

$$[P_2, Q_2] = \frac{1}{2} L_2, \quad [P_2, L_2] = -\frac{1}{2} Q_2, \quad [P_2, Q_3] = L_3,$$

$$[P_2, L_3] = -Q_3, \quad [Q_1, L_1] = j_1^2 P_1, \quad [Q_3, L_3] = j_2^2 P_2,$$

$$[Q_2, L_2] = j_1^2 j_2^2 (P_1 + P_2), \quad [Q_1, L_2] = -\frac{1}{2} j_1^2 Q_3, \tag{3.91}$$

$$[Q_2, L_3] = \frac{1}{2} j_2^2 Q_1, \quad [L_1, Q_2] = \frac{1}{2} j_1^2 Q_3, \quad [Q_3, L_1] = \frac{1}{2} Q_2,$$

$$[L_2, Q_3] = -\frac{1}{2} j_2^2 Q_1, \quad [Q_1, L_3] = -\frac{1}{2} Q_2, \quad [Q_1, Q_3] = \frac{1}{2} L_2,$$

$$[Q_1, Q_2] = \frac{1}{2} j_1^2 L_3, \quad [Q_2, Q_3] = \frac{1}{2} j_2^2 L_1, \quad [L_1, L_2] = \frac{1}{2} j_1^2 L_3,$$

$$[L_1, L_3] = -\frac{1}{2} L_2, \quad [L_2, L_3] = \frac{1}{2} j_2^2 L_1.$$

It is well known [Barut and Raczka (1977)] that the structure of group (algebra) is changed under contraction. Let $j_1 = \iota_1$, then the simple classical algebra su_3 contracts into the semidirect sum of subalgebras $su_3(\iota_1, j_2) = T \oplus u_2(j_2)$, where $T = \{Q_1, L_1, Q_2, L_2\}$ is the commutative ideal and the subalgebra $u_2(j_2) = \{P_1, P_2, Q_3, L_3\}$ is the Lie algebra of the unitary group in the complex Cayley–Klein space $\mathbf{C}_2(j_2)$. From Eq. (3.91) we conclude for $j_1 = \iota_1$ that $[T, u_2(j_2)] \subset T$ as it must be for a semidirect sum. The structure of $SU_3(\iota_1, j_2)$ is the semidirect product $SU_3(\iota_1, j_2) = e^T \otimes U_2(j_2)$. Such groups are called inhomogeneous unitary groups [Perroud (1983)].

From Eq. (3.69) we have the general element of the algebra $su_3(j)$ in the form

$$Z(\mathbf{r}, \mathbf{s}, \mathbf{w}, j) = \sum_{k=1}^{3} (r_k Q_k + s_k L_k) + w_1 P_1 + w_2 P_2$$

$$= \frac{1}{2} \begin{pmatrix} iw_1 & -j_1^2 \zeta_1^* & -j_1^2 j_2^2 \zeta_2^* \\ \zeta_1 & i(w_2 - w_1) & -j_2^2 \zeta_3^* \\ \zeta_2 & \zeta_3 & -iw_2 \end{pmatrix}, \qquad (3.92)$$

where $\zeta_k = s_k + ir_k, k = 1, 2, 3$ and ζ_k^* is the complex conjugate. The finite group transformation matrix $\Xi(\zeta, \mathbf{w}, j)$ is obtained from Eq. (3.92) by the exponential map (3.68). We shall find the matrix Ξ by the Cayley-Hamilton theorem [Korn and Korn (1961)]. The characteristic equation $\det(Z - \lambda E_3) = 0$ of the matrix Z is the following cubic equation:

$$\lambda^3 + p\lambda + q = 0,$$
$$p = w_1^2 - w_1 w_2 + w_2^2 + |\zeta|^2(j),$$
$$q = -iw_1 w_2(w_2 - w_1) + iw_2 j_1^2 |\zeta_1|^2 - i(w_2 - w_1) j_1^2 j_2^2 |\zeta_2|^2$$
$$- iw_1 j_2^2 |\zeta_3|^2 + 2i j_1^2 j_2^2 \mathrm{Im} \zeta_1 \zeta_2^* \zeta_3, \qquad (3.93)$$

where

$$|\zeta|^2(j) = j_1^2 |\zeta_1|^2 + j_1^2 j_2^2 |\zeta_2|^2 + j_2^2 |\zeta_3|^2. \qquad (3.94)$$

The roots of Eq. (3.93) are as follows [Barut and Raczka (1977)]:

$$\lambda_k = \left(-\frac{q}{2} + \sqrt{\left(\frac{q}{2}\right)^2 + \left(\frac{p}{3}\right)^3} \right)^{1/3}$$

$$+ \left(-\frac{q}{2} - \sqrt{\left(\frac{q}{2}\right)^2 + \left(\frac{p}{3}\right)^3} \right)^{1/3} = \lambda_k' + \lambda_k'', \qquad (3.95)$$

where $\lambda_k' + \lambda_k'' = -p/3$ and $\lambda_k,\ k = 1, 2, 3$ are three distinct cube roots. Then by the Cayley–Hamilton theorem we obtain

$$\Xi(\zeta, \mathbf{w}, j) = A \cdot E_3 - B \cdot Z + C \cdot Z^2, \qquad (3.96)$$

where

$$Z^2 = \frac{1}{4}\begin{pmatrix} -w_1^2 - j_1^2|\zeta_1|^2 - j_1^2 j_2^2|\zeta_2|^2 & -j_1^2(iw_2\zeta_1^* + j_2^2\zeta_2^*\zeta_3) \\ iw_2\zeta_1 - j_2^2\zeta_2\zeta_3^* & -(w_2 - w_2)^2 - j_1^2|\zeta_1|^2 - j_2^2|\zeta_3|^2 \\ -i(w_2 - w_1)\zeta_2 + \zeta_1\zeta_3 & -iw_1\zeta_3 - j_1^2\zeta_1^*\zeta_2 \end{pmatrix}$$

$$\begin{pmatrix} j_1^2 j_2^2(i(w_2 - w_1)\zeta_2^* + \zeta_1^*\zeta_3^*) \\ j_2^2(iw_1\zeta_3^* - j_1^2\zeta_1\zeta_2^*) \\ -w_2^2 - j_1^2 j_2^2|\zeta_2|^2 - j_2^2|\zeta_3|^2 \end{pmatrix}, \qquad (3.97)$$

and the functions A, B, C are expressed in the following way:

$$A = D^{-1}[e^{\lambda_1}\lambda_2\lambda_3(\lambda_2 - \lambda_3) - e^{\lambda_2}\lambda_1\lambda_3(\lambda_1 - \lambda_3) + e^{\lambda_3}\lambda_1\lambda_2(\lambda_1 - \lambda_2)],$$
$$B = D^{-1}[e^{\lambda_1}(\lambda_2^2 - \lambda_3^2) - e^{\lambda_2}(\lambda_1^2 - \lambda_3^2) + e^{\lambda_3}(\lambda_1^2 - \lambda_2^2)],$$
$$C = D^{-1}[e^{\lambda_1}(\lambda_2 - \lambda_3) - e^{\lambda_2}(\lambda_1 - \lambda_3) + e^{\lambda_3}(\lambda_1 - \lambda_2)], \qquad (3.98)$$
$$D = (\lambda_1 - \lambda_2)(\lambda_1 - \lambda_3)(\lambda_2 - \lambda_3).$$

From $\text{Tr} Z = 0$ we conclude that $\det \Xi = 1$ and $\Xi^{-1}(\zeta, \mathbf{w}, j) = \Xi(-\zeta, -\mathbf{w}, j)$.

We shall only discuss contractions of the special unitary group SU_3, i.e., $j_1 = 1, \iota_1, j_2 = 1, \iota_2$. Then $\psi\hat{\mathbf{a}}^* = \hat{\mathbf{a}}^+$, $\psi\hat{\mathbf{a}} = \hat{\mathbf{a}}$ in Eq. (3.70) and the Jordan–Schwinger representation of the generators (3.90) is given by the operators

$$\hat{P}_1(j) = \frac{i}{2}(\hat{a}_0^+\hat{a}_0 - \hat{a}_1^+\hat{a}_1), \quad \hat{P}_2(j) = \frac{i}{2}(\hat{a}_1^+\hat{a}_1 - \hat{a}_2^+\hat{a}_2),$$
$$\hat{Q}_1(j) = \frac{i}{2}(\hat{a}_1^+\hat{a}_0 + j_1^2\hat{a}_0^+\hat{a}_1), \quad \hat{L}_1(j) = \frac{1}{2}(\hat{a}_1^+\hat{a}_0 - j_1^2\hat{a}_0^+\hat{a}_1),$$
$$\hat{Q}_2(j) = \frac{i}{2}(\hat{a}_2^+\hat{a}_0 + j_1^2 j_2^2\hat{a}_0^+\hat{a}_2), \quad \hat{L}_2(j) = \frac{1}{2}(\hat{a}_2^+\hat{a}_0 - j_1^2 j_2^2\hat{a}_0^+\hat{a}_2),$$
$$\hat{Q}_3(j) = \frac{i}{2}(\hat{a}_2^+\hat{a}_1 + j_2^2\hat{a}_0^+\hat{a}_2), \quad \hat{L}_3(j) = \frac{1}{2}(\hat{a}_2^+\hat{a}_1 - j_2^2\hat{a}_1^+\hat{a}_2).$$
$$(3.99)$$

These operators satisfy the commutation relations (3.91). The general element of $su_3(j)$ is represented by the following operator:

$$\hat{Z}(\zeta, \mathbf{w}, j) = \sum_{k=1}^{3}(r_k\hat{Q}_k(j) + s_k\hat{L}_k(j)) + w_1\hat{P}_1(j) + w_2\hat{P}_2(j)$$

$$= \frac{1}{2}\left\{ iw_1\hat{a}_0^+\hat{a}_0 + i(w_2 - w_1)\hat{a}_1^+\hat{a}_1 - iw_2\hat{a}_2^+\hat{a}_2 + \right.$$

$$\left. + \sum_{k=1}^{3}\zeta_k\hat{a}_k^+\hat{a}_0 - j_1^2\zeta_1^*\hat{a}_0^+\hat{a}_1 - j_1^2 j_2^2\zeta_2^*\hat{a}_0^+\hat{a}_2 - j_2^2\zeta_3^*\hat{a}_1^+\hat{a}_2 \right\}.$$

$$(3.100)$$

The kernel of the finite group transformation operator (3.71) of $SU_3(j)$ in a coherent state basis is given by Eq. (3.72), using Eqs. (3.96)–(3.98). We shall not write this kernel. Instead, we write out only the kernel associated to the operator of the group $SU_3(\iota_1, \iota_2)$ obtained from SU_3 by two-dimensional contraction, as follows:

$$
U(\alpha^*, \beta, \mathbf{w}, \iota_1, \iota_2) = \exp\Big\{ A' \sum_{k=1}^{3} \alpha_k^* \beta_k + \frac{i}{2} B' w_1 (\alpha_1^* \beta_1 - \alpha_0^* \beta_0)
$$

$$
+ \frac{i}{2} B' w_2 (\alpha_2^* \beta_2 - \alpha_1^* \beta_1) - \frac{1}{4} C' w_1^2 (\alpha_0^* \beta_0 + \alpha_1^* \beta_1)
$$

$$
- \frac{1}{4} C' w_2^2 (\alpha_1^* \beta_1 + \alpha_2^* \beta_2) + \frac{1}{2} C' w_1 w_2 \alpha_1^* \beta_1
$$

$$
+ \frac{1}{2} \Big(\frac{i}{2} w_2 C' - B' \Big) \zeta_1 \alpha_1^* \beta_0 - \frac{1}{2} \Big[\frac{i}{2} (w_2 - w_1) C' + B' \Big]
$$

$$
\times \zeta_2 \alpha_2^* \beta_0 - \frac{1}{2} \Big(\frac{i}{2} w_1 C' + B' \Big) \zeta_3 \alpha_2^* \beta_1 + \frac{1}{4} C' \zeta_1 \zeta_3 \alpha_2^* \beta_0 \Big\},
$$

$$
(3.101)
$$

where the functions A', B', C' are given by Eqs. (3.98) and $\lambda_1, \lambda_2, \lambda_3$ are the roots of Eq. (3.93) with the following coefficients: $p = w_1^2 - w_1 w_2 + w_2^2$, $q = -iw_1 w_2 (w_2 - w_1)$.

We have regarded the unitary Cayley–Klein groups $SU_{n+1}(j)$ as the groups of motion (except for translations) of the complex Cayley–Klein spaces $\mathbf{C}_{n+1}(j)$. The groups $SU_{n+1}(j)$ have been obtained from the classical group SU_{n+1} by contractions and analytical continuations of the group parameters. It has been shown that all these groups are described in a unified way by introducing n parameters $j = (j_1, j_2, \ldots, j_n)$, each of which are equal to the real, dual, or imaginary units. We have built the Jordan–Schwinger representation of the group under consideration. The only contractions of the Jordan–Schwinger representations permit the unified description. In the case of analytical continuations, each of the Jordan–Schwinger representation is built in a particular way.

3.5 The Jordan–Schwinger representations of the symplectic Cayley–Klein groups

In this section we discuss the case of the symplectic Cayley–Klein groups. The symplectic groups and their representations are used in different branches of physics [Biedenharn and Louck (1981)], [Barut and Raczka

(1977)]. The unitary representations of the symplectic $Sp(n, R)$ and pseu-dosymplectic $Sp(p, q)$ groups have been regarded in Refs. [Pajas and Raczka (1968)], [Pajas (1969)]. The oscillator representation for the orthogonal and symplectic groups was discussed by Lohe and Hurst [Lohe and Hurst (1971)].

3.5.1 *The symplectic group* Sp_n

First, we briefly review the necessary information about the symplectic group. As it is well known [Barut and Raczka (1977)], the symplectic group Sp_n includes all transformations of $2n$-dimensional space $\mathbf{R}_n \times \mathbf{R}_n$ under which the following bilinear form is invariant:

$$Sp_n: \quad \mathbf{R}_n \times \mathbf{R}_n \to \mathbf{R}_n \times \mathbf{R}_n,$$

$$[\mathbf{x}, \mathbf{y}] = x_1 y_{-1} - x_{-1} y_1 + \sum_{k=2}^{n}(x_k y_{-k} - x_{-k} y_k), \quad (3.102)$$

where $\{x_k, x_{-k}\}$ is the Cartesian coordinates in $\mathbf{R}_n \times \mathbf{R}_n$. In the space of infinitely differentiable functions $f : \mathbf{R}_n \times \mathbf{R}_n \to \mathbf{R}$ the group Sp_n acts as $g : f(\mathbf{x}, \mathbf{y}) \to f(g\mathbf{x}, g\mathbf{y})$ and their generators are in the form

$$\tilde{X}_{\alpha\beta}(\mathbf{x}) = x_\alpha \partial_\beta - \epsilon_\alpha \epsilon_\beta x_{-\beta} \partial_{-\alpha}, \quad (3.103)$$

where $\alpha, \beta = \pm 1, \pm 2, \ldots, \pm n$, $\partial_\alpha = \partial/\partial x_\alpha$, $\epsilon_\alpha = \text{sign}\,\alpha$, i.e., $\epsilon_\alpha = 1$ for $\alpha > 0$, $\epsilon_\alpha = -1$ for $\alpha < 0$ and $\epsilon_\alpha = 0$ for $\alpha = 0$. The generators (3.103) are not independent. They satisfy the symmetry condition

$$\tilde{X}_{\alpha,\beta} = -\epsilon_\alpha \epsilon_\beta \tilde{X}_{-\beta,-\alpha}. \quad (3.104)$$

Then, as independent generators we choose the following $n(2n + 1)$ generators:

$$\tilde{X}_{\mu\mu}(\mathbf{x}) = x_\mu \partial_\mu - x_{-\mu}\partial_{-\mu}, \quad \mu = 1, 2, \ldots, n,$$

$$\tilde{X}_{\mu,-\mu}(\mathbf{x}) = 2x_\mu \partial_{-\mu}, \quad \mu = \pm 1, \pm 2, \ldots, \pm n,$$

$$\tilde{X}_{\nu\mu}(\mathbf{x}) = x_\nu \partial_\mu - \epsilon_\mu x_{-\mu}\partial_{-\nu},$$

$$\tilde{X}_{\mu\nu}(\mathbf{x}) = x_\mu \partial_\nu - \epsilon_\mu x_{-\nu}\partial_{-\mu}, \quad \nu = 2, 3, \ldots, n, \quad |\mu| < \nu. \quad (3.105)$$

The generators (3.103) of Sp_n satisfy the commutation relations

$$[\tilde{X}_{\alpha\beta}, \tilde{X}_{\alpha'\beta'}] = \delta_{\alpha'\beta}\tilde{X}_{\alpha\beta'} - \delta_{\alpha\beta'}\tilde{X}_{\alpha'\beta} + \epsilon_\alpha \epsilon_\beta \delta_{-\beta',\beta}\tilde{X}_{\alpha',-\alpha}$$

$$+ \epsilon_\beta \epsilon_{\alpha'}\delta_{\alpha',-\alpha}\tilde{X}_{-\beta,\beta'}. \quad (3.106)$$

We may write the generators (3.103), (3.105) in the matrix form using the relation

$$\tilde{X}_{\alpha\beta}(\mathbf{x}) = \partial \tilde{X}_{\alpha\beta}\mathbf{x}, \tag{3.107}$$

where $\partial = (\partial_1, \partial_2, \ldots, \partial_n, \partial_{-1}, \partial_{-2}, \ldots, \partial_{-n})$ is the row matrix, $\mathbf{x} = (x_1, x_2, \ldots, x_n, x_{-1}, x_{-2}, \ldots, x_{-n})^T$ is the column matrix, and the product in Eq. (3.107) is the ordinary matrix product. Then the generators $\tilde{X}_{\alpha\beta}$ are a $2n$-dimensional matrix. The independent generators (3.105) are characterized by the nonzero matrix elements as follows:

$$\begin{aligned}
(\tilde{X}_{\mu\mu})_{\mu\mu} &= 1, \quad (\tilde{X}_{\mu\mu})_{-\mu,-\mu} = -1, \quad \mu = 1, 2, \ldots, n, \\
(\tilde{X}_{\mu,-\mu})_{-\mu,\mu} &= 2, \quad \mu = \pm 1, \pm 2, \ldots, \pm n, \\
(\tilde{X}_{\nu\mu})_{\mu\nu} &= 1, \quad (\tilde{X}_{\nu\mu})_{-\nu,-\mu} = -\epsilon_\mu, \\
(\tilde{X}_{\mu\nu})_{\nu\mu} &= 1, \quad (\tilde{X}_{\mu\nu})_{-\mu,-\nu} = -\epsilon_\mu, \quad \nu = 2, 3, \ldots, n, \ |\mu| < \nu.
\end{aligned} \tag{3.108}$$

3.5.2 The symplectic Cayley–Klein groups $Sp_n(j)$

To obtain the symplectic Cayley–Klein groups, we regard the map

$$\begin{aligned}
\psi &: \mathbf{R}_n \to \mathbf{R}_n(j), \\
\psi x_1^* &= x_1, \quad \psi x_k^* = (1, k) x_k, \quad k = 2, \ldots, n,
\end{aligned} \tag{3.109}$$

where $j = (j_2, j_3, \ldots, j_n)$ and each of the parameters j_m may be equal to the real unit 1 or to the Clifford dual unit or to the imaginary unit i. We agree throughout the section that $(a, b) = \prod_{m=a}^{b} j_m = 1$ for $a > b$. Some applications of the dual numbers in geometry can be found in [Rosenfeld (1955); Yaglom (1963)].

The symplectic Cayley–Klein group $Sp_n(j)$ is defined as the transformation group of $2n$-dimensional space $\mathbf{R}_n(j) \times \mathbf{R}_n(j)$ leaving invariant the following bilinear form:

$$\begin{aligned}
Sp_n(j) &: \mathbf{R}_n(j) \times \mathbf{R}_n(j) \to \mathbf{R}_n(j) \times \mathbf{R}_n(j), \\
[\mathbf{x}, \mathbf{y}] &= x_1 y_{-1} - x_{-1} y_1 + \sum_{k=2}^{n} (1, k)^2 (x_k y_{-k} - x_{-k} y_k).
\end{aligned} \tag{3.110}$$

In accordance with our approach, we obtain the generators of $Sp_n(j)$ by transformations of the generators of Sp_n. From the definition of the generator

$$X(\mathbf{x}) = \sum_{k=-n}^{n} \left. \frac{x_k'}{\partial a} \right|_{a=0} \partial_k,$$

where $\mathbf{x}' = g(a)\mathbf{x}$, $g(0) = 1$, $g(a) \in Sp_n(j)$, $\mathbf{x} \in \mathbf{R}_n(j) \times \mathbf{R}_n(j)$, using the map (3.109), we have the transformation law in the form

$$X_{\alpha\beta}(\mathbf{x}) = (\min\{|\alpha|, |\beta|\}, \max\{|\alpha|, |\beta|\})\tilde{X}_{\alpha\beta}(\psi\mathbf{x}^*). \qquad (3.111)$$

Here $\psi\mathbf{x}^*$ is given by Eq. (3.109), where the upper limit of the production is equal to $|k|$ for negative k. The generators

$$\tilde{X}_{\alpha\beta}(\psi\mathbf{x}^*) = (\min\{|\alpha|, |\beta|\}, \max\{|\alpha|, |\beta|\})^{sign(|\alpha|-|\beta|)} x_\alpha \partial_\beta$$
$$-(\min\{|\alpha|, |\beta|\}, \max\{|\alpha|, |\beta|\})^{-sign(|\alpha|-|\beta|)} \epsilon_\alpha \epsilon_\beta x_{-\beta} \partial_{-\alpha}$$
$$(3.112)$$

are the Wigner–Inönü singularly transformed generators (when some parameters j_k are equal to the dual units). Using Eqs. (3.111) and (3.112) we obtain the generators $X_{\alpha\beta}(\mathbf{x})$ of $Sp_n(j)$ in the form

$$X_{\alpha\beta}(\mathbf{x}) = (\min\{|\alpha|, |\beta|\}, \max\{|\alpha|, |\beta|\})^{1+sign(|\alpha|-|\beta|)} x_\alpha \partial_\beta$$
$$-(\min\{|\alpha|, |\beta|\}, \max\{|\alpha|, |\beta|\})^{1-sign(|\alpha|-|\beta|)} \epsilon_\alpha \epsilon_\beta x_{-\beta} \partial_{-\alpha}.$$
$$(3.113)$$

They also satisfy the symmetry condition (3.104). The independent generators of $Sp_n(j)$ are obtained from Eq. (3.105) by transformations (3.111) or directly from Eq. (3.113), and are given as follows:

$$X_{\mu\mu}(\mathbf{x}) = x_\mu \partial_\mu - x_{-\mu} \partial_{-\mu}, \quad \mu = 1, 2, \dots, n,$$
$$X_{\mu,-\mu}(\mathbf{x}) = 2x_\mu \partial_{-\mu}, \quad \mu = \pm 1, \pm 2, \dots, \pm n,$$
$$X_{\nu\mu}(\mathbf{x}) = (|\mu|, \nu)^2 x_\nu \partial_\mu - \epsilon_\mu x_{-\mu} \partial_{-\nu}, \qquad (3.114)$$
$$X_{\mu\nu}(\mathbf{x}) = x_\mu \partial_\nu - (|\mu|, \nu)^2 \epsilon_\mu x_{-\nu} \partial_{-\mu}, \quad \nu = 2, 3, \dots, n, \quad |\mu| < \nu.$$

From Eq. (3.111) we have

$$\tilde{X}_{\alpha\beta}(\psi\mathbf{x}^*) = (\min\{|\alpha|, |\beta|\}, \max\{|\alpha|, |\beta|\})^{-1} X_{\alpha\beta}(\mathbf{x}).$$

Substituting in Eq. (3.106) the generators $\tilde{X}_{\alpha\beta}$ for their expressions by $X_{\alpha\beta}$ we immediately find the commutation relations of $Sp_n(j)$

$$[X_{\alpha\beta}, X_{\alpha'\beta'}] = (q_1, p_1)(q_1', p_1') \{ (q_2, p_2)^{-1} \delta_{\alpha'\beta} X_{\alpha\beta'}$$
$$-(q_2', p_2')^{-1} \delta_{\alpha\beta'} X_{\alpha'\beta} + (q_3, p_3)^{-1} \epsilon_\alpha \epsilon_\beta \delta_{-\beta',\beta} X_{\alpha',-\alpha}$$
$$+(q_3', p_3')^{-1} \epsilon_\beta \epsilon_{\alpha'} \delta_{\alpha',-\alpha} X_{-\beta,\beta'} \} \qquad (3.115)$$

where $p_1 = \min\{|\alpha|, |\beta|\}$, $q_1 = \max\{|\alpha|, |\beta|\}$, $p_1' = \min\{|\alpha'|, |\beta'|\}$, $q_1' = \max\{|\alpha'|, |\beta'|\}$, $p_2 = \min\{|\alpha|, |\beta'|\}$, $q_2 = \max\{|\alpha|, |\beta'|\}$, $p_2' = \min\{|\alpha'|, |\beta|\}$, $q_2' = \max\{|\alpha'|, |\beta|\}$, $p_3 = \min\{|\alpha|, |\alpha'|\}$, $q_3 = \max\{|\alpha|, |\alpha'|\}$, $p_3' = \min\{|\beta|, |\beta'|\}$, $q_3' = \max\{|\beta|, |\beta'|\}$. At last the independent matrix generators of $Sp_n(j)$ are obtained from Eq. (3.114) using Eq. (3.107). Their nonzero matrix elements are as follows:

$$
\begin{aligned}
(X_{\mu\mu})_{\mu\mu} &= 1, \quad (X_{\mu\mu})_{-\mu,-\mu} = -1, \quad \mu = 1, 2, \ldots, n, \\
(X_{\mu,-\mu})_{-\mu,\mu} &= 2, \quad \mu = \pm 1, \pm 2, \ldots, \pm n, \\
(X_{\nu\mu})_{\mu\nu} &= (|\mu|, \nu)^2, \quad (X_{\nu\mu})_{-\nu,-\mu} = -\epsilon_\mu, \quad (X_{\mu\nu})_{\nu\mu} = 1, \\
(X_{\mu\nu})_{-\mu,-\nu} &= -\epsilon_\mu(|\mu|, \nu)^2, \quad \nu = 2, 3, \ldots, n, \ |\mu| < \nu.
\end{aligned}
\tag{3.116}
$$

The general element $Z(\mathbf{r}, j) = \sum r_{\alpha\beta} X_{\alpha\beta}(j)$ of the algebra $sp_n(j)$, where the sum is doing over all independent generators, is mapped by exponent into the finite group transformation $\Xi(\mathbf{r}, j) = \exp Z(\mathbf{r}, j)$.

The Jordan–Schwinger representation of $Sp_n(j)$ is provided by the operators [cf. with Eq. (3.23)]

$$
\hat{X}_{\alpha\beta}(j) = \psi \hat{\mathbf{a}}^+ X_{\alpha\beta}(j) \psi \hat{\mathbf{a}},
\tag{3.117}
$$

where

$$
\begin{aligned}
\psi \hat{\mathbf{a}}^+ &= \left(\hat{a}_k^+(1, k)^{-1}, \hat{a}_{-k}^+(1, |k|)^{-1} \right), \quad k = 1, 2, \ldots, n, \\
\psi \hat{\mathbf{a}} &= (\hat{a}_k(1, k), \hat{a}_{-k}(1, |k|)).
\end{aligned}
\tag{3.118}
$$

Here ψ is identical, when $j_m = 1, \iota_m$, and for the imaginary values of parameters j we use the well known properties of the annihilation and creation operators: $i\hat{a}_{\pm k} = \hat{a}_{\pm k}^+$, $i\hat{a}_{\pm k}^+ = \epsilon \hat{a}_{\pm k}$. Then Eq. (3.118) may be written in the form

$$
\begin{pmatrix} \psi \hat{\mathbf{a}} \\ \psi \hat{\mathbf{a}}^+ \end{pmatrix} = \begin{pmatrix} \psi_1(j) & -\psi_2(j) \\ \epsilon \psi_2(j) & \psi_1(j) \end{pmatrix} \begin{pmatrix} \hat{\mathbf{a}} \\ \hat{\mathbf{a}}^+ \end{pmatrix} = \Psi^{-1}(j) \begin{pmatrix} \hat{\mathbf{a}} \\ \hat{\mathbf{a}}^+ \end{pmatrix},
\tag{3.119}
$$

where $\psi_1(j), \psi_2(j)$ are $2n$-dimensional diagonal matrices with the following nonzero matrix elements: $(\psi_1)_{\pm 1, \pm 1} = 1$, $(\psi_1)_{\pm k, \pm k} = \pm 1$, if $(1, |k|) = \pm \gamma$ and γ is a positive real or dual number, $(\psi_1)_{\pm k, \pm k} = 0$ otherwise; $(\psi_2)_{\pm 1, \pm 1} = 0$, $(\psi_2)_{\pm k, \pm k} = 0$, if $(\psi_1)_{\pm k, \pm k} = \pm 1$ and $(\psi_2)_{\pm k, \pm k} = \mp 1$, if $(1, |k|) = \pm i\gamma$. The $4n$-dimensional matrix $\Psi(j)$ has the property $\Psi(j) = (\Psi^{-1}(j))^T$.

The finite group transformation is represented by the operator

$$
\hat{U}_q(\mathbf{r}, j) = \exp\{-\hat{Z}(\mathbf{r}, j)\} = \exp\left\{-\sum r_{\alpha\beta} \hat{X}_{\alpha\beta}(j)\right\}.
\tag{3.120}
$$

The kernel of the operator $\hat{U}_g(\mathbf{r}, j)$ in a coherent state basis is obtained in an analogous fashion to the case of orthogonal or unitary groups. When the Cayley–Klein groups $Sp_n(j)$ are obtained from Sp_n only by contractions ($j_k = 1, \iota_k$, $k = 2, 3, \ldots, n$) the kernel is given by

$$U(\alpha^*, \beta, \mathbf{r}, j) = \exp\{\alpha^* \Xi(-\mathbf{r}, j)\beta\}, \tag{3.121}$$

[see Eq. (3.29)], where $\alpha^* = (\alpha_1^*, \ldots \alpha_n^*, \alpha_{-1}^*, \ldots, \alpha_{-n}^*)$, $\beta = (\beta_1, \ldots, \beta_n, \beta_{-1}, \ldots, \beta_{-n})$. When the groups $Sp_n(j)$ are obtained from Sp_n by both contractions and analytical continuations the kernel is given by Eq. (3.7), where the matrices ξ, η, η_1 are expressed through the matrix $\Xi(\mathbf{r}, j)$ by Eqs. (3.32) with $\psi_1(j)$ and $\psi_2(j)$ as in Eq. (3.119).

We shall discuss in detail the groups Sp_1 and $Sp_2(j_2)$. Only for these groups we are able to obtain the explicit form of the finite group transformation matrix $\Xi(\mathbf{r}, j)$.

3.5.3   *Representations of Sp_1 group*

The simplest symplectic group Sp_1 is the transformation group of the two dimensional space $\mathbf{R}_1 \times \mathbf{R}_1$ which preserves the invariance of the bilinear form $[\mathbf{x}, \mathbf{y}] = x_1 y_{-1} - x_{-1} y_1$. The three independent generators of Sp_1 are given by Eq. (3.108) in the form

$$X_{11} = \begin{pmatrix} 1 & 0 \\ 0 & -1 \end{pmatrix}, \quad X_{1,-1} = \begin{pmatrix} 0 & 0 \\ 2 & 0 \end{pmatrix}, \quad X_{-1,1} = \begin{pmatrix} 0 & 2 \\ 0 & 0 \end{pmatrix}, \tag{3.122}$$

and satisfy the commutation relations

$$[X_{11}, X_{1,-1}] = -2X_{1,-1}, \quad [X_{11}, X_{-1,1}] = 2X_{-1,1},$$
$$[X_{1,-1}, X_{-1,1}] = -4X_{11}. \tag{3.123}$$

The general element of the algebra sp_1 is given by the matrix

$$X(r_1, \mathbf{s}) = r_1 X_{11} + s_1 X_{1,-1} + s_2 X_{-1,1} = \begin{pmatrix} r_1 & 2s_2 \\ 2s_1 & -r_2 \end{pmatrix}, \tag{3.124}$$

and the finite transformation matrix Ξ is

$$\Xi(r_1, \mathbf{s}) = \exp X(r_1, \mathbf{s}) = E \cosh p + X \frac{\sinh p}{p}$$

$$= \begin{pmatrix} \cosh p + r_1 \frac{\sinh p}{p} & 2s_2 \frac{\sinh p}{p} \\ 2s_1 \frac{\sinh p}{p} & \cosh p - r_1 \frac{\sinh p}{p} \end{pmatrix} \tag{3.125}$$

where $p^2 = r_1^2 + 4s_1 s_2$.

Following (3.117) the Jordan–Schwinger representation of Sp_1 is provided by the operators

$$\hat{X}_{11} = \hat{a}_1^+ \hat{a}_1 - \hat{a}_{-1}^+ \hat{a}_{-1}, \quad \hat{X}_{1,-1} = 2\hat{a}_{-1}^+ \hat{a}_1, \quad \hat{X}_{-1,1} = 2\hat{a}_1^+ \hat{a}_{-1}. \quad (3.126)$$

The kernel of the operator $\hat{U}_g(r_1, \mathbf{s}) = \exp\{-\hat{X}(r_1, \mathbf{s})\}$ in a coherent state basis is given by Eqs. (3.125), (3.121) or explicitly by

$$U(\alpha^*, \beta, r_1, \mathbf{s}) = \exp\left\{ (\alpha_1^* \beta_1 + \alpha_{-1}^* \beta_{-1}) \cosh p \right.$$

$$\left. - \frac{\sinh p}{p} [r_1(\alpha_1^* \beta_1 - \alpha_{-1}^* \beta_{-1}) + 2s_1\alpha_{-1}^* \beta_1 + 2s_2\alpha_1^* \beta_{-1}] \right\}.$$

$$(3.127)$$

3.5.4 *Representations of $Sp_2(j_2)$ groups*

The group $Sp_2(j_2)$ consists of all transformations of $\mathbf{R}_2(j_2) \times \mathbf{R}_2(j_2)$ leaving invariant the bilinear form $[\mathbf{x}, \mathbf{y}] = x_1 y_{-1} - x_{-1} y_1 + j_2^2 (x_2 y_{-2} - x_{-2} y_2)$. The ten independent matrix generators of $Sp_2(j_2)$ are given by Eq. (3.116) as follows:

$$X_{11} = \begin{pmatrix} 1 & 0 & 0 & 0 \\ 0 & 0 & 0 & 0 \\ 0 & 0 & -1 & 0 \\ 0 & 0 & 0 & 0 \end{pmatrix}, \quad X_{1,-1} = \begin{pmatrix} 0 & 0 & 0 & 0 \\ 0 & 0 & 0 & 0 \\ 2 & 0 & 0 & 0 \\ 0 & 0 & 0 & 0 \end{pmatrix},$$

$$X_{-1,1} = \begin{pmatrix} 0 & 0 & 2 & 0 \\ 0 & 0 & 0 & 0 \\ 0 & 0 & 0 & 0 \\ 0 & 0 & 0 & 0 \end{pmatrix}, \quad X_{22} = \begin{pmatrix} 0 & 0 & 0 & 0 \\ 0 & 1 & 0 & 0 \\ 0 & 0 & 0 & 0 \\ 0 & 0 & 0 & -1 \end{pmatrix},$$

$$X_{2,-2} = \begin{pmatrix} 0 & 0 & 0 & 0 \\ 0 & 0 & 0 & 0 \\ 0 & 0 & 0 & 0 \\ 0 & 2 & 0 & 0 \end{pmatrix}, \quad X_{-2,2} = \begin{pmatrix} 0 & 0 & 0 & 0 \\ 0 & 0 & 0 & 2 \\ 0 & 0 & 0 & 0 \\ 0 & 0 & 0 & 0 \end{pmatrix},$$

$$X_{12}(j_2) = \begin{pmatrix} 0 & 0 & 0 & 0 \\ 1 & 0 & 0 & 0 \\ 0 & 0 & 0 & -j_2^2 \\ 0 & 0 & 0 & 0 \end{pmatrix}, \quad X_{21}(j_2) = \begin{pmatrix} 0 & j_2^2 & 0 & 0 \\ 0 & 0 & 0 & 0 \\ 0 & 0 & 0 & 0 \\ 0 & 0 & -1 & 0 \end{pmatrix},$$

$$X_{2,-1}(j_2) = \begin{pmatrix} 0 & 0 & 0 & 0 \\ 0 & 0 & 0 & 0 \\ 0 & j_2^2 & 0 & 0 \\ 1 & 0 & 0 & 0 \end{pmatrix}, \quad X_{-1,2}(j_2) = \begin{pmatrix} 0 & 0 & 0 & j_2^2 \\ 0 & 0 & 1 & 0 \\ 0 & 0 & 0 & 0 \\ 0 & 0 & 0 & 0 \end{pmatrix}.$$

$$(3.128)$$

Let us observe that each set of generators $A_1 = \{X_{11}, X_{1,-1}, X_{-1,1}\}$ and $A_2 = \{X_{22}, X_{2,-2}, X_{-2,2}\}$ form the subalgebra of $sp_2(j_2)$, isomorphic to the algebra sp_1, with the commutators (3.123) and $[A_1, A_2] = 0$. Then $A = A_1 \oplus A_2$ is the subalgebra of $sp_2(j_2)$. The other nonzero commutation relations of $sp_2(j_2)$ are obtained from Eq. (3.115) in the form

$$[X_{11}, X_{12}] = -X_{12}, \quad [X_{11}, X_{21}] = X_{21},$$
$$[X_{11}, X_{2,-1}] = -X_{2,-1}, \quad [X_{11}, X_{-1,2}] = X_{-1,2},$$
$$[X_{1,-1}, X_{21}] = 2X_{2,-1}, \quad [X_{1,-1}, X_{-1,2}] = -2X_{12},$$
$$[X_{-1,1}, X_{12}] = -2X_{-1,2}, \quad [X_{-1,1}, X_{2,-1}] = 2X_{21},$$

$$(3.129)$$

$$[X_{22}, X_{12}] = X_{12}, \quad [X_{22}, X_{21}] = -X_{21},$$
$$[X_{22}, X_{2,-1}] = -X_{2,-1}, \quad [X_{22}, X_{-1,2}] = X_{-1,2},$$
$$[X_{2,-2}, X_{12}] = 2X_{2,-1}, \quad [X_{2,-2}, X_{-1,2}] = -2X_{21},$$
$$[X_{-2,2}, X_{21}] = -2X_{-1,2}, \quad [X_{-2,2}, X_{2,-1}] = 2X_{12},$$
$$[X_{12}, X_{21}] = j_2^2(X_{22} - X_{11}), \quad [X_{12}, X_{2,-1}] = -j_2^2 X_{1,-1},$$
$$[X_{12}, X_{-1,2}] = j_2^2 X_{-2,2}, \quad [X_{21}, X_{2,-1}] = -j_2^2 X_{2,-2},$$

$$(3.130)$$

$$[X_{21}, X_{-1,2}] = j_2^2 X_{-1,1}, \quad [X_{2,-1}, X_{-1,2}] = -j_2^2(X_{11} + X_{22}). \quad (3.131)$$

For $j_2 = \iota_2$ we conclude from Eq. (3.131) that the set $T = \{X_{12}, X_{21}, X_{2,-1}, X_{-1,2}\}$ is the commutative ideal of the algebra $sp_2(\iota_2)$, since it follows from Eqs. (3.129), (3.130) that $[T, A_1] \subset T$ and $[T, A_2] \subset T$, i.e., $[T, A] \subset T$. Then the structure of the algebra $sp_2(\iota_2)$ is the semidirect sum $sp_2(\iota) = T \,\dot{\oplus}\, A = T \,\dot{\oplus}\, (A_1 \oplus A_2)$ and for the contracted group $Sp_2(\iota_2)$ we obtain the structure of the semidirect product $Sp_2(\iota_2) = e^T \,\dot{\otimes}\, (Sp_1 \times Sp_1)$. It must be emphasized that unlike the case of unitary groups, the contracted symplectic groups are not the inhomogeneous groups in the sense of [Perroud (1983)] and [Chaichian, Demichev and Nelipa (1983)] since the latter have the structure $R_{2n} \,\dot{\otimes}\, Sp_n$.

The general element of the algebra $sp_2(j_2)$ may be written in the form

$$X(j_2) = r_1 X_{11} + r_2 X_{22} + s_1 X_{1,-1} + s_2 X_{2,-2} + w_1 X_{-1,1}$$

$$+ w_2 X_{-2,2} + u_1 X_{12} + u_2 X_{-1,2} + v_1 X_{21} + v_2 X_{2,-1}$$

$$= \begin{pmatrix} r_1 & j_2^2 v_1 & 2w_1 & j_2^2 u_2 \\ u_1 & r_2 & u_2 & 2w_2 \\ 2s_1 & j_2^2 v_2 & -r_1 & -j_2^2 u_1 \\ v_2 & 2s_2 & -v_1 & -r_2 \end{pmatrix}. \tag{3.132}$$

We shall obtain the finite group transformation matrix $\Xi(j_2)$ by the Cayley-Hamilton theorem [Korn and Korn (1961)]. The characteristic equation $\det(X(j_2) - \lambda \cdot E) = 0$ of the matrix $X(j_2)$ is the following biquadratic equation:

$$\lambda^4 - p\lambda^2 + q = 0,$$

$$p = \mathbf{r}^2 + 4(\mathbf{s}, \mathbf{w}) + 2j_2^2(\mathbf{u}, \mathbf{v}),$$

$$q = (r_1^2 + 4s_1 w_1)(r_2^2 + 4s_2 w_2) + j_2^4(\mathbf{u}, \mathbf{v})^2 + 2j_2^2(r_1 r_2 u_2 v_2 - r_1 r_2 u_1 v_1$$

$$- 2r_1 u_1 u_2 s_2 - 2r_1 v_1 v_2 w_2 - 2r_2 u_2 v_1 s_1 - 2r_2 u_1 v_2 w_1 + 2v_1^2 s_1 w_2$$

$$+ 2u_1^2 w_1 s_2 - 2v_2^2 w_1 w_2 - 2u_2^2 s_1 s_2). \tag{3.133}$$

Let $\lambda^2 = z$, then $z_{1,2} = \frac{1}{2}(p \pm \sqrt{p^2 - 4q})$ and the roots of Eq. (3.133) are as follows: $\lambda_{1,2} = \pm\sqrt{z_1}, \lambda_{3,4} = \pm\sqrt{z_2}$. The matrix $\Xi(j_2) = \exp X(j_2)$ is given by the Cayley-Hamilton theorem in the form

$$\Xi(j_2) = \frac{1}{\sqrt{p^2 - 4q}} \left\{ E\left(z_1 \cosh\sqrt{z_2} - z_2 \cosh\sqrt{z_1} \right) \right.$$

$$+ X(j_2) \left(\frac{z_1 \sinh\sqrt{z_2}}{\sqrt{z_2}} - \frac{z_2 \sinh\sqrt{z_1}}{\sqrt{z_1}} \right) + X^2(j_2)$$

$$\left. \times \left(\cosh\sqrt{z_1} - \cosh\sqrt{z_2} \right) + X^3(j_2) \left(\frac{\sinh\sqrt{z_1}}{\sqrt{z_1}} - \frac{\sinh\sqrt{z_2}}{\sqrt{z_2}} \right) \right\}, \tag{3.134}$$

where the matrices $X^2(j_2)$ and $X^3(j_2)$ are characterized by the matrix elements

$$(X^2(j_2))_{kk} = (X^2(j_2))_{-k,-k} = r_k^2 + 4s_k w_k + j_2^2(\mathbf{u}, \mathbf{v}), \quad k = 1, 2,$$

$$(X^2(j_2))_{21} = u_1(r_1 + r_2) + 2(s_1 u_2 + v_2 w_2),$$

$$(X^2(j_2))_{-1,-2} = j_2^2 (X^2(j_2))_{21},$$

$$(X^2(j_2))_{-2,-1} = v_1(r_1 + r_2) + 2(u_2 s_2 + v_2 w_1),$$

$$(X^2(j_2))_{12} = j_2^2 (X^2(j_2))_{-2,-1},$$

$$(X^2(j_2))_{-2,1} = v_2(r_1 - r_2) + 2(u_1 s_2 - v_1 s_1),$$

$$(X^2(j_2))_{-1,2} = -j_2^2 (X^2(j_2))_{-2,1},$$

$$(X^2(j_2))_{2,-1} = u_2(r_2 - r_1) + 2(u_1 w_1 - v_1 w_2),$$

$$(X^2(j_2))_{1,-2} = -j_2^2 (X^2(j_2))_{2,-1},$$

$$(X^2(j_2))_{1,-1} = (X^2(j_2))_{-1,1} = (X^2(j_2))_{2,-2} = (X^2(j_2))_{-2,2} = 0.$$

$$(3.135)$$

$$(X^3(j_2))_{11} = -(X^3(j_2))_{-1,-1} = r_1[r_1^2 + 4s_1 w_1$$

$$+2j_2^2(\mathbf{u},\mathbf{v})] + j_2^2[r_2(u_1 v_1 - u_2 v_2) + 2(u_1 u_2 s_2 + v_1 v_2 w_2)],$$

$$(X^3(j_2))_{22} = -(X^3(j_2))_{-2,-2} = r_2[r_2^2 + 4s_2 w_2$$

$$+2j_2^2(\mathbf{u},\mathbf{v})] + j_2^2[r_1(u_1 v_1 - u_2 v_2) + 2(u_1 v_2 w_1 + u_2 v_1 s_1)],$$

$$(X^3(j_2))_{21} = u_1[r_1^2 + r_1 r_2 + r_2^2 + 4(\mathbf{s},\mathbf{w}) + j_2^2(\mathbf{u},\mathbf{v})]$$

$$+2(r_1 v_2 w_2 + r_2 u_2 s_1 - 2v_1 s_1 w_2),$$

$$(X^3(j_2))_{-2,1} = v_2[r_1^2 - r_1 r_2 + r_2^2 + 4(\mathbf{s},\mathbf{w}) + j_2^2(\mathbf{u},\mathbf{v})]$$

$$+2(r_1 u_1 s_2 + r_2 v_2 s_1 + 2u_2 s_1 s_2),$$

$$(X^3(j_2))_{2,-1} = u_2[r_1^2 - r_1 r_2 + r_2^2 + 4(\mathbf{s},\mathbf{w}) + j_2^2(\mathbf{u},\mathbf{v})]$$

$$+2(r_1 v_1 w_2 + r_2 u_1 w_1 + 2v_2 w_1 w_2),$$

$$(X^3(j_2))_{-2,-1} = -v_1[r_1^2 + r_1 r_2 + r_2^2 + 4(\mathbf{s},\mathbf{w}) + j_2^2(\mathbf{u},\mathbf{v})]$$

$$-2(r_1 u_2 s_2 + r_2 v_2 w_1 - 2u_1 s_2 w_1),$$

$$(X^3(j_2))_{-1,1} = 2s_1[r_1^2 + 4s_1 w_1 + 2j_2^2(\mathbf{u},\mathbf{v})]$$

$$+2j_2^2(r_2 u_1 v_2 + v_2^2 w_2 - u_1^2 s_2),$$

$$(X^3(j_2))_{-2,2} = 2s_2[r_2^2 + 4s_2w_2 + 2j_2^2(\mathbf{u},\mathbf{v})]$$

$$+2j_2^2(r_1v_1v_2 + v_2^2w_1 - v_1^2s_1),$$

$$(X^3(j_2))_{1,-1} = 2w_1[r_1^2 + 4s_1w_1 + 2j_2^2(\mathbf{u},\mathbf{v})]$$

$$+2j_2^2(r_2u_2v_1 + u_2^2s_2 - v_1^2w_2),$$

$$(X^3(j_2))_{2,-2} = 2w_2[r_2^2 + 4s_2w_2 + 2j_2^2(\mathbf{u},\mathbf{v})]$$

$$+2j_2^2(r_1u_1u_2 + u_2^2s_1 - u_1^2w_1),$$

$$(X^3(j_2))_{-1,-2} = -j_2^2(X^3(j_2))_{21},$$

$$(X^3(j_2))_{12} = -j_2^2(X^3(j_2))_{-2,-1},$$

$$(X^3(j_2))_{1,-2} = j_2^2(X^3(j_2))_{2,-1},$$

$$(X^3(j_2))_{-1,2} = j_2^2(X^3(j_2))_{-2,1}. \tag{3.136}$$

According to Eq. (3.121) the matrix $\Xi(j_2)$ is only needed to obtain the kernel of the finite transformation operator of $Sp_2(j_2), j_2 = 1, \iota$ in the Jordan–Schwinger representation. We shall not write out this kernel.

3.6 Concluding remarks

In this chapter using the well-developed theory of quantum systems with quadratic behavior in creation and annihilation boson or fermion Hamiltonian operators we have built the Jordan–Schwinger representations of the groups under consideration. The matrix elements of the Jordan–Schwinger representation of the finite group transformation operator in a Glauber coherent state basis are obtained with help of the finite transformation matrix Ξ. In the case of contractions these matrix elements are completely defined by the matrix Ξ and in the case of both contractions and analytical continuations we have introduced the map $\Psi(j)$ that transforms the matrix Ψ into the matrix ξ, η, η_1. The last case includes a nonlinear operation of obtaining the inverse matrix ξ^{-1} and therefore for the matrix elements we have the more complicated equations, as in the first case. For the boson representations the matrix elements under consideration are the generating functions for the matrix elements in discrete Fock bases. The last matrix elements are expressed in terms of Hermite polynomials of several variables with zero arguments.

The Jordan–Schwinger representations of groups are closely connected with the properties of stationary quantum systems whose Hamiltonians are quadratic in creation and annihilation operators. The replacement of a group parameter $\mathbf{r}$ by $(i/\hbar)t\mathbf{r}$, where t is a time variable, transforms the matrix elements of the finite group operator into the Greens function of corresponding quantum systems. Thus, the unified description of the Jordan–Schwinger representations of the Cayley–Klein groups gives us opportunity to investigate the sets of stationary quantum systems.

Chapter 4

The Gel'fand–Tsetlin representations of Cayley–Klein algebras

In this chapter the contractions of Gel'fand–Tsetlin irreducible representations of unitary and orthogonal algebras, which are especially convenient for applications in quantum physics, are studied. The most general contractions, which give representations with nonzero spectrum of all Casimir operators, are considered in detail. Representations of algebras with the structures of semidirect sum are obtained from well known irreducible representations for classical algebras by the method of transition. When algebras contracted on different parameters are isomorphic, we get representations in different (discrete and continuous) bases.

4.1 Representations of unitary algebras $u(2; j_1)$ and $su(2; j_1)$

4.1.1 Finite-dimensional irreducible representations of algebras $u(2)$ and $su(2)$

These representations have been described by I.M. Gel'fand and M.L. Tsetlin [Gel'fand and Tsetlin (1950)]. They are realized in the space with orthogonal basis, determined by a pattern with integer-valued components

$$|m^*\rangle = \begin{vmatrix} m_{12}^* & & m_{22}^* \\ & m_{11}^* & \end{vmatrix}\bigg\rangle, \quad m_{12}^* \geq m_{11}^* \geq m_{22}^*, \tag{4.1}$$

by operators

$$E_{11}^*|m^*\rangle = m_{11}^*|m^*\rangle = A_{11}^*|m^*\rangle,$$
$$E_{22}^*|m^*\rangle = (m_{12}^* + m_{22}^* - m_{11}^*)|m^*\rangle = A_{00}^*|m^*\rangle,$$

$$E_{21}^*|m^*\rangle = \sqrt{(m_{12}^* - m_{11}^* + 1)(m_{11}^* - m_{22}^*)}|m_{11}^* - 1\rangle = A_{01}^*|m\rangle,$$

$$E_{12}^*|m^*\rangle = \sqrt{(m_{12}^* - m_{11}^*)(m_{11}^* + 1 - m_{22}^*)}|m_{11}^* + 1\rangle = A_{10}^*|m\rangle, \qquad (4.2)$$

where $|m_{11}^* \pm 1\rangle$ denotes the pattern (4.1) with the component m_{11}^* substituted for $m_{11}^* \pm 1$. Let us change the standard notations of generators E_{kr} for new notations $A_{n-k,n-r}$, $n = 2$, consistent with the notations of section 1.3. The irreducible representation is completely determined by the components $m_{12}^*, m_{22}^*, m_{12}^* \geq m_{22}^*$ of the upper row in (4.1) (components of the highest weight).

As it is known, Casimir operators are proportional to the unit operators on the space of irreducible representation. The spectrum of Casimir operators for classical groups was obtained in [Perelomov and Popov (1966a,b)] and for semisimple groups in [Leznov, Malkin and Man'ko (1977)]. For algebra $u(2)$ it is as follows:

$$C_1^* = m_{12}^* + m_{22}^*, \quad C_2^* = m_{12}^{*2} + m_{22}^{*2} + m_{12}^* - m_{22}^*. \qquad (4.3)$$

Let us remind that the asterisk marks the quantities referring to the classical groups (algebras).

In the representation space there is the vector of the highest weight z_h, described by the pattern (4.1) for $m_{11}^* = m_{12}^*$. Acting on it, the raising operator A_{10}^* gives zero: $A_{10}^* z_h = 0$ and the lowering operator A_{01}^* makes the value $m_{11}^* = m_{12}^*$ less by one: $A_{01}^* z_h = \sqrt{m_{12}^* - m_{22}^*}|m_{12}^* - 1\rangle$. Consequently, applying A_{01}^* to z_h, we come to the vector of the lowest weight z_l, described by the pattern (4.1) for $m_{11}^* = m_{22}^*$. Acting on z_l the lowering operator gives zero: $A_{01}^* z_l = 0$. The irreducible representation is finite-dimensional, and this fact is reflected in the inequalities (4.1), which are satisfied by the component m_{11}^* of the pattern.

The unitarity condition for representations of the algebra $u(2)$ is equivalent to the following relations for the operators (4.2): $A_{kk}^* = \bar{A}_{kk}^*$, $k = 0, 1$, $A_{01}^* = \bar{A}_{10}^*$, where the bar means the complex conjugation. For the matrix elements the unitarity condition can be written as follows

$$\langle m^*|A_{kk}^*|m^*\rangle = \overline{\langle m^*|A_{kk}^*|m^*\rangle}, \ k = 0, 1,$$

$$\langle m^*|A_{11}^*|m^*\rangle = \overline{\langle m^*|A_{11}^*|m^*\rangle}, \qquad (4.4)$$

$$\langle m_{11}^* - 1|A_{01}^*|m^*\rangle = \overline{\langle m^*|A_{10}^*|m_{11}^* - 1\rangle}.$$

The representations of special unitary algebra $su(2)$ result from the representations of $u(2)$ for $m_{12}^* = l^*$, $m_{22}^* = -l^*$, $m_{11}^* \equiv m^*$, $|m^*| \leq l^*$

and are given by operators $J_3^* = A_{11}^*$, $J_-^* = \frac{1}{\sqrt{2}}A_{01}^*$, $J_+^* = \frac{1}{\sqrt{2}}A_{10}^*$ with the commutation relations

$$[J_3^*, J_\pm^*] = \pm J_\pm^*, \quad [J_+^*, J_-^*] = J_3^*. \tag{4.5}$$

Eigenvalues of Casimir operator C_1^* vanish, and for the second order Casimir operator are as follows $C_2^* = l^*(l^* + 1)$.

4.1.2   Transition to the representations of algebras $u(2; j_1)$ and $su(2; j_1)$

Under transition from the algebra $u(2)$ to the algebra $u(2; j_1)$ the generators A_{00}^*, A_{11}^* and the Casimir operator C_1^* remain unchanged, and the generators A_{01}^*, A_{10}^* and the Casimir operator C_2^* are transformed as follows:

$$A_{01} = j_1 A_{01}^*(\rightarrow), \quad A_{10} = j_1 A_{10}^*(\rightarrow), \quad C_2(j_1) = j_1^2 C_2^*(\rightarrow), \tag{4.6}$$

where $A_{01}^*(\rightarrow)$, $A_{10}^*(\rightarrow)$ are singularly transformed (for nilpotent value of parameter $j_1 = \iota_1$) generators of the initial algebra $u(2)$. The question of interest now is how do we set this transformation for the irreducible representation (4.2) of the algebra $u(2)$. Let us give the transformation of the component of pattern (4.1) as follows:

$$m_{12} = j_1 m_{12}^*, \quad m_{22} = j_1 m_{22}^*, \quad m_{11} = m_{11}^*. \tag{4.7}$$

Then, taking into account of (4.6), the representation generators (4.2) can be written as

$$A_{00}|m\rangle = \left(\frac{m_{12} + m_{22}}{j_1} - m_{11}\right)|m\rangle, \quad A_{11}|m\rangle = m_{11}|m\rangle,$$
$$A_{01}|m\rangle = \sqrt{(m_{12} - j_1 m_{11} + j_1)(j_1 m_{11} - m_{22})}|m_{11} - 1\rangle, \tag{4.8}$$
$$A_{10}|m\rangle = \sqrt{(m_{12} - j_1 m_{11})(j_1 m_{11} + j_1 - m_{22})}|m_{11} + 1\rangle,$$

and the spectrum of Casimir operators

$$C_1(j_1) = A_{00} + A_{11},$$
$$C_2(j_1) = A_{01}A_{10} + A_{10}A_{01} + j_1^2(A_{00}^2 + A_{11}^2) \tag{4.9}$$

are

$$C_1(j_1) = \frac{1}{j_1}(m_{12} + m_{22}),$$
$$C_2(j_1) = m_{12}^2 + m_{22}^2 + j_1(m_{12} - m_{22}), \tag{4.10}$$

where $|m\rangle$ means the following pattern

$$|m\rangle = \left|\begin{array}{cc} m_{12} & m_{22} \\ & m_{11} \end{array}\right\rangle. \tag{4.11}$$

The inequalities (4.1) for components can be formally written as

$$\frac{m_{12}}{j_1} \geq m_{11} \geq \frac{m_{22}}{j_1}, \quad \frac{m_{12}}{j_1} \geq \frac{m_{22}}{j_1}. \tag{4.12}$$

To reveal the sense of these inequalities for $j_1 \neq 1$, we shall discuss the action of raising operator A_{10} on the vector of the "highest weight" z_h, described by pattern (4.11) for $m_{11} = m_{12}$, and the lowering operator A_{01} on the vector of the "lowest weight" z_l, described by pattern (4.11) for $m_{11} = m_{22}$. We obtain

$$\begin{aligned} A_{10}z_h &= \sqrt{m_{12}(1 - j_1)(j_1 m_{12} + j_1 - m_{22})}|m_{12} + 1\rangle, \\ A_{10}z_l &= \sqrt{(m_{12} - j_1 m_{22} + j_1)m_{22}(j_1 - 1)}|m_{22} - 1\rangle. \end{aligned} \tag{4.13}$$

It can be seen from this that for $j_1 = \iota_{1,i}$ these expressions differ from zero. Therefore, the space of representation is infinite-dimensional, and the integer-valued component m_{11}, which numbers the basis vectors, is varying from $-\infty$ to ∞. Thus, the formal inequalities (4.12) for $j_1 = \iota_{1,i}$ are interpreted as $\infty > m_{11} > -\infty$ and $m_{12} \geq m_{22}$.

The form (4.7) of the transformation of Gel'fand–Tsetlin pattern is chosen in such a way that the second order Casimir operator would differ from zero and does not contain indeterminate expressions for j_1.

Under transition from algebra $su(2)$ to algebra $su(2; j_1)$ the generators and components of Gel'fand–Tsetlin patterns are transformed as follows:

$$J_\pm = j_1 J_\pm^*(\rightarrow), \quad J_3 = J_3^*(\rightarrow), \quad l = j_1 l^*, \quad m = m^*. \tag{4.14}$$

We have in result

$$\begin{aligned} J_\pm|l, m\rangle &= \alpha^\pm(m)|l, m \pm 1\rangle, \quad J_3|l, m\rangle = m|l, m\rangle, \\ \alpha^\pm(m) &= \frac{1}{\sqrt{2}}\sqrt{(l \mp j_1 m)(l \pm j_1 m + j_1)}, \end{aligned} \tag{4.15}$$

which satisfy the commutation relations

$$[J_3, J_\pm] = \pm J_\pm, \quad [J_+, J_-] = j_1^2 J_3. \tag{4.16}$$

Casimir operator has the spectrum $C_2(j_1) = l(l + j_1)$ on the irreducible representation.

4.1.3   *Contractions of irreducible representations*

For $j_1 = \iota_1$ the operator A_{00} contains the term $(m_{12} + m_{22})/\iota_1$, which is in general indeterminate, if its numerator is a real, complex number. This term is determinate if its numerator is proportional to the nilpotent number ι_1, i.e. $m_{12} + m_{22} = \iota_1 s$, where $s \in \mathbb{R}, \mathbb{C}$. The requirement of unitarity for the operator A_{00} is given by $s \in \mathbb{R}$. Thus, in order that the operators (4.8) determine the representation of the algebra $u(2; \iota_1)$, it is necessary to choose the components m_{12}, m_{22} of the pattern (4.11) in the form

$$m_{12} = k + \iota_1 \frac{s}{2}, \quad m_{22} = -k + \iota_1 \frac{s}{2}, \quad s \in \mathbb{R}, \tag{4.17}$$

where k, generally speaking, is a complex number. The pattern (base vector) (4.11) for nilpotent values of components is determined by the expansion in series

$$|m\rangle = \left|\begin{matrix} k + \iota_1 \frac{s}{2} & & -k + \iota_1 \frac{s}{2} \\ & m_{11} & \end{matrix}\right\rangle = |\tilde{m}\rangle + \iota_1 \frac{s}{2}\left(|\tilde{m}\rangle'_{12} + |\tilde{m}\rangle'_{22}\right),$$

$$|\tilde{m}\rangle = \left|\begin{matrix} k & & -k \\ & m_{11} & \end{matrix}\right\rangle, \quad |\tilde{m}\rangle'_{12} = \left(\frac{\partial|m\rangle}{\partial m_{12}}\right)_{|m_{12}=k,\, m_{22}=-k}, \tag{4.18}$$

and similar expressions are valid for $|\tilde{m}\rangle'_{22}$. The initial patterns (4.1) are normalized to unit: $\langle m'^{*}|m^{*}\rangle = \delta_{m'^{*}_{12}, m^{*}_{12}} \delta_{m'^{*}_{22}, m^{*}_{22}} \delta_{m'^{*}_{11}, m^{*}_{11}}$. The patterns (4.11) for the continuous values of components are normalized to delta-function. In particular, for $|\tilde{m}\rangle$ we have normalization to the squared delta-function

$$\langle \tilde{m}'|\tilde{m}\rangle = \delta^2(k' - k)\delta_{m'_{11}, m_{11}}. \tag{4.19}$$

Substituting (4.17), (4.18) in formulas of section 4.1.2 and comparing complex parts, we obtain the representation operators of algebra $u(2; \iota_1)$ (the nilpotent parts are omitted):

$$A_{00}|\tilde{m}\rangle = (s - m_{11})|\tilde{m}\rangle, \quad A_{11}|\tilde{m}\rangle = m_{11}|\tilde{m}\rangle,$$

$$A_{01}|\tilde{m}\rangle = k|\tilde{m}_{11} - 1\rangle, \quad A_{10}|\tilde{m}\rangle = k|\tilde{m}_{11} + 1\rangle. \tag{4.20}$$

The requirement of unitarity (4.4) for operators A_{01}, A_{10} gives $k = \bar{k}$, i.e. k is a real number, the inequality $m_{12} \geq m_{22}$ gives for the real parts $k \geq -k$, i.e. $k \geq 0$, the component m_{11} is integer-valued and changes in the range $-\infty < m_{11} < \infty$. The eigenvalues of Casimir operators (4.9) on the irreducible representations of algebra $u(2; \iota_1)$ are obtained according to (4.10) and are equal

$$C_1(\iota_1) = s, \quad C_2(\iota_1) = 2k^2. \tag{4.21}$$

They are independent and differ from zero. As in the case of the initial algebra $u(2)$, the irreducible representations of the contracted algebra $u(2; \iota_1)$ are completely determined by the upper row of the scheme, i.e. by parameters $k \geq 0$, $s \in \mathbb{R}$. The results (4.18), (4.19) coincide with the corresponding formulas in the paper [Chakrabarti (1968)] for the case of inhomogeneous algebra $iu(1)$.

The requirement of determinacy of the spectrum of operator $C_2(\iota_1)$ are not only met by the transformation (4.7) of the components of Gel'fand-Tsetlin pattern but also the transformation $m_{12} = j_1 m_{12}^*$, $m_{22} = m_{22}^*$, $m_{11} = m_{11}^*$ as well. Here generator $A_{00}|m\rangle = (m_{12}/\iota_1 + m_{22} - m_{11})|m\rangle$ is determined only for $m_{12} = \iota_1 p$, $p \in \mathbb{R}$, but then $C_1(\iota_1) = p + m_{22} \neq 0$, and $C_2(\iota_1) = \iota_1^2[m_{22}(m_{22} - 1) + m_{12}^2/\iota_1^2 + m_{12}/\iota_1] = m_{12}^2 + \iota_1 m_{12} = (\iota_1 p)^2 + \iota_1(\iota_1 p) = 0$. In this case the irreducible representation of the algebra $u(2)$ is contracted to the degenerate representation of algebra $u(2; \iota_1)$, for which $C_1(\iota_1) \neq 0$, and $C_2(\iota_1) = 0$. One cannot transform the components $m_{kr} = m_{kr}^*$ at all. Then under contraction we also obtain the degenerate representation of the algebra $u(2; \iota_1)$ with $C_1(\iota_1) = m_{12} + m_{22} \neq 0$, and $C_2(\iota_1) = 0$. This representation is given by generators A_{00}, A_{11} of the form (4.2), and the generators A_{01} and A_{10} bring $|m\rangle$ to zero: $A_{01}|m\rangle = 0$, $A_{10}|m\rangle = 0$.

We have chosen the transformation (4.7) which under contraction gives the non-degenerated general representation of the algebra $u(2; \iota_1)$ with non-zero spectrum of all Casimir operators. Further, in studying the algebras of higher dimensions, we shall consider just this case.

If one takes $j_1 = \iota_1$ in (4.15), one obtains the infinite-dimensional representation of the contracted algebra $su(2; \iota_1)$, which is realized by the operators

$$J_\pm|l, m\rangle = l|l, m \pm 1\rangle, \quad J_3|l, m\rangle = m|l, m\rangle, \quad l \geq 0, \ m \in \mathbb{Z} \tag{4.22}$$

with commutation relations $[J_3, J_\pm] = \pm J_\pm$, $[J_+, J_-] = 0$, and Casimir operator $C_2(\iota_1) = l^2$.

4.1.3.1 *Analytical continuation of irreducible representations*

As it has been noticed in section 4.1.3, the formulas for the transformation of algebraic quantities, derived from the requirement of the absence of indeterminate expressions for nilpotent values in contraction parameters, are valid for imaginary values of parameters as well. For algebra $u(2; j_1 = i) \equiv u(1, 1)$ this means that $(m_{12} + m_{22})/i = s$. The requirement of the unitarity for A_{00} gives $s \in \mathbb{R}$, i.e. the components m_{12} and m_{22}, in general, are

$$m_{12} = a + i\left(b + \frac{s}{2}\right), \quad m_{22} = -a - i\left(b - \frac{s}{2}\right), \quad a, b, s \in \mathbb{R}. \tag{4.23}$$

Substituting (4.23) in (4.8), (4.10), we get the generators

$$A_{\substack{01\\10}}|m\rangle$$

$$= \sqrt{a^2 - b(b+1) + \left(\frac{s}{2} - m_{11}\right)\left(\frac{s}{2} - m_{11} \pm 1\right) + ia(2b+1)}|m_{11} \mp 1\rangle,$$

$$A_{00}|m\rangle = (s - m_{11})|m\rangle, \quad A_{11}|m\rangle = m_{11}|m\rangle, \tag{4.24}$$

and Casimir operators

$$C_1(i) = s, \quad C_2(i) = 2\left[a^2 - b(b+1) - \left(\frac{s}{2}\right)^2\right] + 2ia(2b+1). \tag{4.25}$$

The relation (4.4) for the operators A_{01}, A_{10}, implied by the requirement of Hermiticity, can be written as follows:

$$\sqrt{a^2 - b(b+1) + \left(\frac{s}{2} - m_{11}\right)\left(\frac{s}{2} - m_{11} + 1\right) + ia(2b+1)}$$

$$= \sqrt{a^2 - b(b+1) + \left(\frac{s}{2} - m_{11}\right)\left(\frac{s}{2} - m_{11} + 1\right) - ia(2b+1)}. \tag{4.26}$$

To satisfy (4.26) for any s, m_{11}, the imaginary part of the radicand must vanish and the real part must be positive. It is possible in two cases: (A) $b = -1/2$, $a \neq 0$; (B) $a = 0$, $-b(b+1) > 0$.

In the case (A) the formulas (4.24) can be rewritten as follows:

$$A_{\substack{01\\10}}|m\rangle = \sqrt{a^2 + \left(m_{11} \mp \frac{1-s}{2}\right)^2}|m_{11} \mp 1\rangle,$$

$$C_1(i) = s, \quad C_2(i) = 2a^2 + \frac{1}{2}(1 - s^2). \tag{4.27}$$

This is the irreducible representation of the continuous series of the algebra $u(1, 1)$. Gel'fand and Graev (1965) used the components $\tilde{m}_{12} = -1/2 + r$, $\tilde{m}_{22} = 1/2 + r$ related to components m_{12}, m_{22} via formulas $m_{12} = i\tilde{m}_{12}$, $m_{22} = i\tilde{m}_{22}$, i.e. $\text{Re}(r) = s/2$, $\text{Im}(r) = -a$.

In the case (B) the relations (4.24) can be rewritten as follows:

$$A_{\substack{01\\10}}|m\rangle = \sqrt{\left(m_{11} - \frac{s \pm 1}{2}\right)^2 - \left(\frac{b+1}{2}\right)^2}\,|m_{11} \mp 1\rangle,$$

$$C_1(i) = s, \quad C_2(i) = -2\left(b(b+1) + \frac{s^2}{4}\right). \tag{4.28}$$

This is the irreducible representation of additional continuous series [Linblad and Nagel (1970)].

There is one more possibility apart from cases (A) and (B). Let the components m_{12}, m_{22} be purely imaginary: $m_{12} = i\hat{m}_{12}$, $m_{22} = i\hat{m}_{22}$, where $\hat{m}_{12}$, $\hat{m}_{22}$ are integers. Then the relations (4.24) can be rewritten as follows:

$$A_{00}|m\rangle = (\hat{m}_{12} + \hat{m}_{22} - m_{11})|m\rangle, \quad A_{11}|m\rangle = m_{11}|m\rangle,$$

$$A_{01}|m\rangle = \sqrt{-(\hat{m}_{12} - m_{11} + 1)(m_{11} - \hat{m}_{22})}\,|m_{11} - 1\rangle,$$

$$A_{10}|m\rangle = \sqrt{-(\hat{m}_{12} - m_{11})(m_{11} + 1 - \hat{m}_{22})}\,|m_{11} + 1\rangle, \tag{4.29}$$

$$C_1(i) = \hat{m}_{12} + \hat{m}_{22}, \quad C_2(i) = -(\hat{m}_{12}^2 + \hat{m}_{22}^2 + \hat{m}_{12} - \hat{m}_{22}).$$

They coincide with (4.2), (4.3), except for the minus sign in the radicand. The requirement of unitarity (4.4) can be reduced to the reality of the root in the expressions for the generators A_{01}, A_{10}, which is possible when one of the factors is negative. As a result, we get two more irreducible representations: (C) $m_{11} > \hat{m}_{12} + 1$; (D) $m_{11} < \hat{m}_{22} - 1$, which are called discrete series. The discrete series of the irreducible representations of pseudounitary algebras $u(p,q)$ are described by Gel'fand and Graev (1965, 1967). The cases (C) and (D) correspond to modified patterns

$$\left|\begin{matrix} \hat{m}_{12} & \hat{m}_{22} \\ m_{11} & \end{matrix}\right\rangle, \quad \left|\begin{matrix} \hat{m}_{12} & \hat{m}_{22} \\ & m_{11} \end{matrix}\right\rangle. \tag{4.30}$$

In the simplest case of algebras $u(2; j_1)$ we have shown in detail how the method of transitions works for irreducible representations. The irreducible representations of algebras $u(2; j_1)$ are given by formulas of section 4.1.2 with additional conditions (4.17) in the case of contraction and (4.23) in the case of analytical continuation to the components of the upper row in Gel'fand–Tsetlin pattern. To obtain unitary representation, it is necessary to check additionally whether the relations (4.4) are satisfied for contracted and analytically continued generators of representation.

4.2 Representations of unitary algebras $u(3; j_1, j_2)$

4.2.1 *Description of representations*

The standard notations of I.M. Gel'fand and M.L. Tsetlin correspond to diminishing chain of subalgebras $u(3) \supset u(2) \supset u(1)$, where $u(3) = \{E_{kr}, k, r = 1, 2, 3\}, u(2) = \{E_{kr}, k, r = 1, 2\}, u(1) = \{E_{11}\}$. To make them consistent with our notations, it is necessary to change index k for index $n - k = 3 - k$, i.e. $E_{kr} = A_{n-k,n-r}$. Doing so, we turn the chain of subalgebras into $u(3; j_1, j_2) \supset u(2; j_2) \supset u(1)$, where $u(3; j) = \{A_{sp}, s, p = 0, 1, 2\}$, $u(2; j_2) = \{A_{sp}, s, p = 1, 2\}, u(1) = \{A_{22}\}$. We leave the component enumeration in Gel'fand–Tsetlin patterns unchanged.

It is well known that to determine representations of algebra $u(3)$ it is sufficient to determine the action of generators E_{pp}, $E_{p,p-1}$, $E_{p-1,p}$, i.e. the generators A_{kk}, $k = 0, 1, 2$, $A_{k+1,k}$, $A_{k,k+1}$, $k = 0, 1$. The remaining generators A_{02}, A_{20} can be found using commutators $A_{02} = [A_{01}, A_{12}]$, $A_{20} = [A_{21}, A_{10}]$. Under transition from $u(3)$ to $u(3; j)$ the generators are transformed as follows: $A_{01} = j_1 A_{01}^*(\rightarrow)$, $A_{12} = j_2 A_{12}^*(\rightarrow)$, $A_{10} = j_1 A_{10}^*(\rightarrow)$, $A_{21} = j_2 A_{21}^*(\rightarrow)$, $A_{kk} = A_{kk}^*(\rightarrow)$. The transformation of the components of Gel'fand–Tsetlin patterns can be defined as follows:

$$m_{13} = j_1 j_2 m_{13}^*, \quad m_{23} = m_{23}^*, \quad m_{33} = j_1 j_2 m_{33}^*,$$
$$m_{12} = j_2 m_{12}^*, \quad m_{22} = j_2 m_{22}^*, \quad m_{11} = m_{11}^*, \tag{4.31}$$

Then the component of the pattern $|m\rangle$ satisfies the inequalities

$$|m\rangle = \left| \begin{matrix} m_{13} & & m_{23} & & m_{33} \\ & m_{12} & & m_{22} & \\ & & m_{11} & & \end{matrix} \right\rangle,$$

$$\frac{m_{13}}{j_1 j_2} \geq m_{23} \geq \frac{m_{33}}{j_1 j_2}, \quad \frac{m_{13}}{j_1 j_2} \geq \frac{m_{12}}{j_2} \geq m_{23}, \tag{4.32}$$

$$m_{23} \geq \frac{m_{22}}{j_2} \geq \frac{m_{33}}{j_1 j_2}, \quad \frac{m_{12}}{j_2} \geq m_{11} \geq \frac{m_{22}}{j_2}.$$

Transforming the known expressions for generators of algebra $u(3)$, we come to the generators of representations of algebra $u(3; j)$:

$$A_{00}|m\rangle = \left(m_{23} + \frac{m_{13} + m_{33}}{j_1 j_2} - \frac{m_{12} + m_{22}}{j_2} \right) |m\rangle,$$

$$A_{01}|m\rangle = \frac{1}{j_2} \left\{ -\frac{(m_{13} - j_1 m_{12} + j_1 j_2)(m_{33} - j_1 m_{12} - j_1 j_2)(j_2 m_{23} - m_{12})}{(m_{22} - m_{12})} \right.$$

$$\times \left.\frac{(j_2 m_{11} - m_{12})}{(m_{22} - m_{12} - j_2)}\right\}^{1/2} |m_{12} - j_2\rangle$$

$$+ \frac{1}{j_2}\left\{\frac{-(m_{13} - j_1 m_{22} + 2j_1 j_2)(m_{33} - j_1 m_{22})(j_2 m_{23} + j_2 - m_{22})}{(m_{12} - m_{22} + 2j_2)}\right.$$

$$\times \left.\frac{(j_2 m_{11} + j_2 - m_{22})}{(m_{12} - m_{22} + j_2)}\right\}^{1/2} |m_{22} - j_2\rangle,$$

$$A_{10}|m\rangle = \frac{1}{j_2}\left\{\frac{-(m_{13} - j_1 m_{12})(m_{33} - j_1 m_{12} - 2j_1 j_2)(j_2 m_{23} - j_2 - m_{12})}{(m_{22} - m_{12} - j_2)}\right.$$

$$\times \left.\frac{(j_2 m_{11} - j_2 - m_{12})}{(m_{22} - m_{12} - 2j_2)}\right\}^{1/2} |m_{12} + j_2\rangle$$

$$+ \frac{1}{j_2}\left\{\frac{-(m_{13} - j_1 m_{22} + j_1 j_2)(m_{33} - j_1 m_{22} - j_1 j_2)(j_2 m_{23} - m_{22})}{(m_{12} - m_{22} + j_2)}\right.$$

$$\times \left.\frac{(j_2 m_{11} - m_{22})}{(m_{12} - m_{22})}\right\}^{1/2} |m_{22} + j_2\rangle,$$

$$A_{02}|m\rangle = \left\{\frac{-(m_{13} - j_1 m_{12} + j_1 j_2)(m_{33} - j_1 m_{12} - j_1 j_2)(j_2 m_{23} - m_{12})}{(m_{22} - m_{12})}\right.$$

$$\times \left.\frac{(m_{22} - j_2 m_{11})}{(m_{22} - m_{12} - j_2)}\right\}^{1/2} \left|\begin{matrix} m_{12} - j_2 \\ m_{11} - 1 \end{matrix}\right\rangle$$

$$+ \left\{\frac{-(m_{13} - j_1 m_{22} + 2j_1 j_2)(m_{33} - j_1 m_{22})(j_2 m_{23} + j_2 - m_{22})}{(m_{12} - m_{22} + 2j_2)}\right.$$

$$\times \left.\frac{(m_{12} - j_2 m_{11} + j_2)}{(m_{12} - m_{22} + j_2)}\right\}^{1/2} \left|\begin{matrix} m_{22} - j_2 \\ m_{11} - 1 \end{matrix}\right\rangle,$$

$$A_{20}|m\rangle = \left\{\frac{-(m_{13} - j_1 m_{12})(m_{33} - j_1 m_{22} - 2j_1 j_2)(j_2 m_{23} - j_2 - m_{12})}{(m_{22} - m_{12} - j_2)}\right.$$

$$\times \left.\frac{(m_{22} - j_2 m_{11} - j_2)}{(m_{22} - m_{12} - 2j_2)}\right\}^{1/2} \left|\begin{matrix} m_{12} + j_2 \\ m_{11} + 1 \end{matrix}\right\rangle$$

$$+ \left\{\frac{-(m_{13} - j_1 m_{22} + j_1 j_2)(m_{33} - j_1 m_{22} - j_1 j_2)(j_2 m_{23} - m_{22})}{(m_{12} - m_{22} + j_2)}\right.$$

$$\times \left.\frac{(m_{12} - j_2 m_{11})}{(m_{12} - m_{22})}\right\}^{1/2} \left|\begin{matrix} m_{22} + j_2 \\ m_{11} + 1 \end{matrix}\right\rangle, \tag{4.33}$$

where $|m_{12} \pm j_2\rangle$ is the pattern (4.32), in which the component m_{12} is substituted for $m_{12} \pm j_2$ and so on. The generators A_{11}, A_{22}, A_{12}, A_{21}, making subalgebra $u(2; j_2)$, are described by (4.8), where each index of the generators must be increased by a unit and the parameter j_1 has to be substituted for the parameter j_2. The generators (4.33) satisfy the commutation relations of algebra $u(3; j)$

$$[A_{kr}, A_{pq}] = \frac{(k, r)(r, q)}{(k, q)} \delta_{pr} A_{kq} - \frac{(k, r)(r, q)}{(p, r)} \delta_{kq} A_{pr}. \qquad (4.34)$$

The unitary algebra $u(3)$ has three Casimir operators, which under transition to algebra $u(3; j)$ are transformed as follows (1.98):

$$C_1(j) = C_1^*(\to), \quad C_2(j) = j_1^2 j_2^2 C_2^*(\to), \quad C_3(j) = j_1^2 j_2^2 C_3^*(\to). \qquad (4.35)$$

The spectrum of Casimir operators in this case is as follows

$$C_1(j) = \frac{m_{13} + m_{33}}{j_1 j_2} + m_{23},$$

$$C_2(j) = m_{13}^2 + m_{33}^2 + j_1^2 j_2^2 m_{23}^2 + 2 j_1 j_2 (m_{13} - m_{33}),$$

$$C_3(j) = \frac{m_{13}^3 + m_{33}^3}{j_1 j_2} + 2(2m_{13}^2 - m_{33}^2) - m_{13} m_{33}$$

$$+ j_1^2 j_2^2 (m_{23}^3 + 2m_{23}^2 - 2m_{23})$$

$$+ j_1 j_2 [2(2m_{13} - m_{33}) - m_{23}(m_{13} + m_{33})]. \qquad (4.36)$$

This naturally raises the question: for what reasons do we choose the transformation rule (4.31) for components of Gel'fand–Tsetlin patterns or the rule (4.7) in the case of algebra $u(2; j_1)$? We choose it in order to make the spectrum of second order Casimir operators different from zero and avoid indeterminate expressions for nilpotent values of parameters j. As $C_2(j) = j_1^2 j_2^2 C_2^*(\to)$ and components m_{13}, m_{23}, m_{33} enter C_2^* quadratically, this requirement gives (4.31). However, the variant (4.31) (we call it basic) is not unique. Two other variants are possible as well: $m_{13} = m_{13}^*$, $m_{23} = j_1 j_2 m_{23}^*$, $m_{33} = j_1 j_2 m_{33}^*$ or $m_{13} = j_1 j_2 m_{13}^*$, $m_{23} = j_1 j_2 m_{23}^*$, $m_{33} = m_{33}^*$, which turn the initial irreducible representation of algebra $u(3)$ into representations of algebra $u(3; j)$ with other (in comparison with basic invariant (4.36)) values of Casimir operators. For example,

$$C_2'(j) = m_{23}^2 + m_{33}^2 + j_1^2 j_2^2 m_{13}(m_{13} + 2) - 2 j_1 j_2 m_{33},$$

$$C''_2(j) = m_{13}^2 + m_{23}^2 + j_1^2 j_2^2 m_{33}(m_{33} - 2) + 2 j_1 j_2 m_{13}. \qquad (4.37)$$

The consideration of these variants of transition for irreducible representations is quite similar to the basic variant, and we skip the corresponding relations.

It will be shown below that the basic transformations (4.31), as well as the other two options, give non-degenerate representations of contracted algebras, all Casimir operators of which are independent and have a nonzero spectrum.

For an interpretation of the formal inequalities (4.32) let us consider the action of raising generator A_{10} on the "highest weight" vector z_h, described by the pattern (4.32) for $m_{11} = m_{12} = m_{13}$, $m_{22} = m_{23}$, and the action of lowering generator A_{01} on the "lowest weight" vector z_l, described by the pattern (4.32) for $m_{11} = m_{22} = m_{33}$, $m_{12} = m_{23}$. Let us write out explicitly only those factors, which vanish for $j_1 = j_2 = 1$. Then

$$A_{10}z_h = \frac{1}{j_2}\sqrt{m_{13}(1-j_1)A}|m_{13}+j_2\rangle + \frac{1}{j_2}\sqrt{m_{23}(j_2-1)B}|m_{23}+j_2\rangle,$$

$$A_{01}z_l = \frac{1}{j_2}\sqrt{m_{23}(j_2-1)C}|m_{23}-j_2\rangle + \frac{1}{j_2}\sqrt{m_{33}(1-j_1)D}|m_{33}-j_2\rangle.$$

$$(4.38)$$

It can be seen from this that for $j_1 \neq 1$, $j_2 = 1$

$$A_{10}z_h = \sqrt{m_{13}(1-j_1)A}|m_{13}+1\rangle \neq 0,$$

which means the absence of an upper bound on m_{12}, and

$$A_{01}z_l = \sqrt{m_{33}(1-j_1)D}|m_{33}-1\rangle \neq 0,$$

that means the absence of a lower bound on m_{22}, i.e. the components of the pattern (4.32) satisfy the inequalities $m_{12} \geq m_{23} \geq m_{22}$. From (4.38) we obtain for $j_1 = 1$, $j_2 \neq 1$

$$A_{10}z_h = j_2^{-1}\sqrt{m_{23}(j_2-1)B}|m_{23}+j_2\rangle \neq 0,$$

implying again the absence of an upper bound on m_{22} and

$$A_{01}z_l = j_2^{-1}\sqrt{m_{23}(j_2-1)C}|m_{23}-j_2\rangle \neq 0,$$

reveals the absence of a lower bound on m_{12}, i.e. the components of the pattern scheme (4.32) satisfy inequalities $m_{13} \geq m_{12}$, $m_{22} \geq m_{33}$, $-\infty < m_{11} < \infty$. At last, we find from (4.38) for $j_1 \neq 1$, $j_2 \neq 1$ that there are no restrictions for components m_{12}, m_{22}, m_{11}. In all cases the inequality $m_{13} \geq m_{33}$ remains valid.

The same inequalities for the components of Gel'fand–Tsetlin pattern can be derived from the formal inequalities (4.32), if one interprets them for j_1, $j_2 \neq 1$ according to the following rules: a) inequality $j^{-1}m \geq m_1$ means the absence of the upper bounds m_1; b) inequality $m_1 \geq j^{-1}m$ means the absence of the lower bounds on m_1; c) inequality $(j_1 j_2)^{-1}m \geq j_1^{-1}m_1$ is equivalent to $j_1^{-1}m \geq m_1$, i.e. common parameters in both parts of the inequality can be canceled out. The same rules are valid for algebras of higher dimensions as well.

Formulas for irreducible representations of algebra $u(3)$ can be obtained from the formulas of this paragraph for $j_1 = j_2 = 1$. The requirement of unitarity for representations of algebra $u(3)$ leads to the following relations for the operators (4.33): $A_{kk} = \bar{A}_{kk}$, $k = 0, 1, 2$, $A_{rp} = \bar{A}_{pr}$, $r, p = 0, 1, 2$. Here the bar means complex conjugation.

4.2.2   *Contraction over the first parameter*

The structure of the contracted unitary algebra is as follows: $u(3; \iota_1, j_2) = T_4 \oplus (u(1) \oplus u(2; j_2))$, where $T_4 = \{A_{01}, A_{10}, A_{02}, A_{20}\}$ is the subalgebra $u(2; j_2) = \{A_{11}, A_{22}, A_{12}, A_{21}\}$ and $u(1) = \{A_{00}\}$. The relations (4.33) for $j_1 = \iota_1$ give

$$A_{00}|m\rangle = \left(\frac{m_{13} + m_{33}}{\iota_1 j_2} + m_{23} - \frac{m_{12} + m_{22}}{j_2} \right) |m\rangle,$$

$$A_{01}|m\rangle = \frac{1}{j_2} \sqrt{-m_{13}m_{33}} \left\{ \sqrt{\frac{(j_2 m_{23} - m_{12})}{(m_{22} - m_{12})} \frac{(j_2 m_{11} - m_{12})}{(m_{22} - m_{12} - j_2)}} |m_{12} - j_2\rangle \right.$$

$$\left. + \sqrt{\frac{(j_2 m_{23} + j_2 - m_{22})}{(m_{12} - m_{22} + 2j_2)} \frac{(j_2 m_{11} + j_2 - m_{22})}{(m_{12} - m_{22} + j_2)}} |m_{22} - j_2\rangle \right\},$$

$$A_{10}|m\rangle = \frac{1}{j_2} \sqrt{-m_{13}m_{33}}$$

$$\times \left\{ \sqrt{\frac{(j_2 m_{23} - j_2 - m_{12})}{(m_{22} - m_{12} - j_2)} \frac{(j_2 m_{11} - j_2 - m_{12})}{(m_{22} - m_{12} - 2j_2)}} |m_{12} + j_2\rangle \right.$$

$$\left. + \sqrt{\frac{(j_2 m_{23} - m_{22})}{(m_{12} - m_{22} + j_2)} \frac{(j_2 m_{11} - m_{22})}{(m_{12} - m_{22})}} |m_{22} + j_2\rangle \right\},$$

$$A_{02}|m\rangle = \sqrt{-m_{13}m_{33}} \left\{ \sqrt{\frac{(j_2m_{23} - m_{12})}{(m_{22} - m_{12})} \frac{(m_{22} - j_2m_{11})}{(m_{22} - m_{12} - j_2)}} \left|\begin{array}{c} m_{12} - j_2 \\ m_{11} - 1 \end{array}\right\rangle\right.$$

$$\left. + \sqrt{\frac{(j_2m_{23} + j_3 - m_{22})}{(m_{12} - m_{22} + 2j_2)} \frac{(m_{12} - j_2m_{11} + j_2)}{(m_{12} - m_{22} + j_2)}} \left|\begin{array}{c} m_{22} - j_2 \\ m_{11} - 1 \end{array}\right\rangle\right\},$$

$$A_{20}|m\rangle = \sqrt{-m_{13}m_{33}}$$

$$\times \left\{ \sqrt{\frac{(j_2m_{23} - j_2 - m_{12})}{(m_{22} - m_{12} - j_2)} \frac{(m_{22} - j_2m_{11} - j_2)}{(m_{22} - m_{12} - 2j_2)}} \left|\begin{array}{c} m_{12} + j_2 \\ m_{11} + 1 \end{array}\right\rangle\right.$$

$$\left. + \sqrt{\frac{(j_2m_{23} - m_{22})}{(m_{12} - m_{22} + j_2)} \frac{(m_{12} - j_2m_{11})}{(m_{12} - m_{22})}} \left|\begin{array}{c} m_{22} + j_2 \\ m_{11} + 1 \end{array}\right\rangle\right\}. \qquad (4.39)$$

Here the nilpotent parts, arising in the expressions for the generators, are omitted, and only real parts are written.

Algebra $u(3; \iota_1, 1)$ is an inhomogeneous algebra $IU(2)$ in notations of Chakrabarti (1968). The requirement of determinacy and unitarity of generator A_{00} gives $(m_{13} + m_{33})/\iota_1 = q \in \mathbb{R}$, i.e.

$$m_{13} = k + \iota_1 \frac{q}{2}, \quad m_{33} = -k + \iota_1 \frac{q}{2}, \quad k, q \in \mathbb{R}, \ k \geq 0. \qquad (4.40)$$

Reality of k follows from the unitary relations for A_{01}, A_{10}, and its positiveness — from the inequality $m_{13} \geq m_{33}$, considered for real parts. Taking into account (4.40), we get $\sqrt{-m_{13}m_{33}} = k$, and the expressions (4.39) for $j_2 = 1$ coincide with the corresponding formulas in Chakrabarti (1968) for $iu(2)$. The integer components of the pattern $|\tilde{m}\rangle$ are interrelated via the inequalities $m_{12} \geq m_{23} \geq m_{22}$, $m_{12} \geq m_{11} \geq m_{22}$, ensued from (4.32) for $j_1 = \iota_1$. The pattern $|\tilde{m}\rangle$ can be obtained from (4.32) for $m_{13} = k$, $m_{33} = -k$. The spectrum of Casimir operators in the given irreducible representation of algebra $u(3; \iota_1, 1)$ can be found from (4.36):

$$C_1(\iota_1, 1) = q + m_{23}, \quad C_2(\iota_1, 1) = 2k^2, \quad C_3(\iota_1, 1) = 3k^2(q + 1). \qquad (4.41)$$

Algebra $su(3, \iota_1, 1)$ differs from algebra $u(3, \iota_1, 1)$ as diagonal operators satisfy the relation $A_{00} + A_{11} + A_{22} = 0$. Acting on the pattern $|\tilde{m}\rangle$, we find the spectrum of Casimir operators

$$C_1(\iota_1, 1) = 0, \quad C_2(\iota_1, 1) = 2k^2, \quad C_3(\iota_1, 1) = 3k^2(1 - m_{23}) \qquad (4.42)$$

of the irreducible representation of algebra $su(3; \iota_1, 1) = T_4 \rightthreetimes u(2)$, generators of which are described by (4.39) for $j_2 = 1$, where it is necessary to put

$$m_{13} = k - \iota_1 \frac{m_{23}}{2}, \quad m_{33} = -k - \iota_1 \frac{m_{23}}{2}, \quad k \geq 0, \quad m_{23} \in \mathbb{Z}. \quad (4.43)$$

Here $\mathbb{Z}$ is a set of integers.

4.2.3 Contraction over the second parameter

The structure of the contracted algebra is as follows: $u(3; j_1, \iota_2) = T_4 \rightthreetimes (u(2; j_1) \oplus u(1))$, where $T_4 = \{A_{12}, A_{21}, A_{02}, A_{20}\}$, $u(2; j_1) = \{A_{00}, A_{11}, A_{01}, A_{10}\}$, $u(1) = \{A_{22}\}$. After the substitution of $j_2 = \iota_2$ in (4.33) the expressions $|m_{12} \pm \iota_2\rangle$ can occur, which according to the general rules of treating functions of nilpotent variable, we expand into the series

$$|m_{12} \pm \iota_2\rangle = |m\rangle \pm \iota_2 |m\rangle'_{12}, \quad |m\rangle'_{12} = \frac{\partial |m\rangle}{\partial m_{12}},$$

$$|m_{22} \pm \iota_2\rangle = |m\rangle \pm \iota_2 |m\rangle'_{22}, \quad |m\rangle'_{22} = \frac{\partial |m\rangle}{\partial m_{22}}. \quad (4.44)$$

Taking this remark into account, for $j_2 = \iota_2$ (4.33) gives the following expressions for the generators:

$$A_{00}|m\rangle = \left(m_{23} + \frac{m_{13} + m_{33}}{\iota_2 j_1} - \frac{m_{12} + m_{22}}{\iota_2} \right) |m\rangle,$$

$$A_{11}|m\rangle = \left(\frac{m_{12} + m_{22}}{\iota_2} - m_{11} \right) |m\rangle, \quad A_{22}|m\rangle = m_{11}|m\rangle,$$

$$A_{12}|m\rangle = \sqrt{-m_{12}m_{22}}|m_{11} - 1\rangle, \quad A_{21}|m\rangle = \sqrt{-m_{12}m_{22}}|m_{11} + 1\rangle,$$

$$A_{01}|m\rangle = \frac{1}{m_{12} - m_{22}} \left\{ \frac{m_{12}a_{12} + m_{22}a_{22}}{\iota_2} |m\rangle \right.$$

$$+ \frac{1}{2a_{12}} \left[j_1 m_{12}(m_{13} - m_{33}) - a_{12}^2 \frac{m_{11} + m_{23} + m_{12}}{m_{12} - m_{22}} \right] |m\rangle$$

$$- \frac{1}{2a_{22}} \left[2j_1 m_{22}(m_{33} - j_1 m_{22}) + a_{22}^2 \left(m_{11} + m_{23} + \frac{2m_{12} + m_{22}}{m_{12} - m_{22}} \right) \right] |m\rangle$$

$$\left. - m_{12}a_{12}|m\rangle'_{12} - m_{22}a_{22}|m\rangle'_{22} \right\},$$

$$A_{10}|m\rangle = \frac{1}{m_{12} - m_{22}} \left\{ \frac{m_{12}a_{12} + m_{22}a_{22}}{\iota_2}|m\rangle \right.$$

$$+ \frac{1}{2a_{12}}\left[2j_1 m_{12}(m_{13} - j_1 m_{12}) - a_{12}^2\left(m_{11} + m_{23} + \frac{m_{12} + 2m_{22}}{m_{12} - m_{22}}\right)\right]|m\rangle$$

$$+ \frac{1}{2a_{22}}\left[j_1 m_{22}(m_{13} - m_{33}) - a_{22}\left(m_{11} + m_{23} + \frac{m_{22}}{m_{12} - m_{22}}\right)\right]|m\rangle$$

$$\left. + m_{12}a_{12}|m\rangle'_{12} + m_{22}a_{22}|m\rangle'_{22} \right\},$$

$$A_{\substack{02 \\ 20}}|m\rangle = \frac{\sqrt{-m_{12}m_{22}}}{m_{12} - m_{22}}(a_{12} + a_{22})|m_{11} \mp 1\rangle,$$

$$a_{12} = \sqrt{-(m_{13} - j_1 m_{12})(m_{33} - j_1 m_{12})},$$

$$a_{22} = \sqrt{-(m_{13} - j_1 m_{22})(m_{33} - j_1 m_{22})}, \qquad (4.45)$$

where only real parts are written.

The requirement for determinacy of generators A_{00}, A_{11} together with the condition of their Hermiticity brings to the components of the patterns

$$m_{13} = k + \iota_2 j_1 \frac{q}{2}, \quad m_{33} = -k + \iota_2 j_1 \frac{q}{2}, \quad k \geq 0, \; q \in \mathbb{R},$$
$$m_{12} = r + \iota_2 \frac{s}{2}, \quad m_{33} = -r + \iota_2 \frac{s}{2}, \quad r \geq 0, \; s \in \mathbb{R}. \qquad (4.46)$$

Substituting the components (4.46) in (4.32), we get

$$|m\rangle = |\tilde{m}\rangle + \iota_2 j_1 q \frac{1}{2}(|\tilde{m}\rangle'_{13} + |\tilde{m}\rangle'_{33}) + \iota_2 s \frac{1}{2}(|\tilde{m}\rangle'_{12} + |\tilde{m}\rangle'_{22}),$$

$$|\tilde{m}\rangle = \left| \begin{array}{ccc} k & & -k \\ & m_{23} & \\ & r & -r \\ & & m_{11} \end{array} \right\rangle, \quad m_{11}, m_{23} \in \mathbb{Z}. \qquad (4.47)$$

For classical unitary algebras, Gel'fand–Tsetlin patterns $|m\rangle$ with the integer components enumerate basis vectors normalized to a unit in finite-dimensional representation space. Under contraction and analytical continuations a part of the components of patterns $|m\rangle$ takes continuous values. In this case the basis vectors in the infinite-dimensional representation space for contracted or analytically continued algebras, corresponding to such patterns, are understood to be generalized functions, orthogonal as before, but normalized to the delta-function. In particular, for $|\tilde{m}\rangle$ we get

$$\langle \tilde{m}'|\tilde{m}\rangle = \delta^2(k'-k)\delta^2(r'-r)\delta_{m'_{23},m_{23}}\delta_{m'_{11},m_{11}}, \tag{4.48}$$

where the squared delta-functions occur due to the fact that r and k enter the components of the pattern twice. Let us compare this to [Celeghini and Tarlini (1981a,b, 1982)] in the case of contractions and to [Kuriyan, Mukunda and Sudarshan (1968a,b); Mukunda (1967)] in the case of analytical continuations.

Substituting (4.46), (4.47) in (4.45), we obtain the generators of representation of algebra $u(3; j_1, \iota_2)$:

$$A_{00}|\tilde{m}\rangle = (m_{23}+q-s)|\tilde{m}\rangle, \quad A_{11}|\tilde{m}\rangle = (s-m_{11})|\tilde{m}\rangle, \quad A_{22}|\tilde{m}\rangle = m_{11}|\tilde{m}\rangle,$$

$$A_{\substack{12\\21}}|\tilde{m}\rangle = r|\widetilde{m_{11}\mp 1}\rangle, \quad A_{\substack{02\\20}}|\tilde{m}\rangle = \sqrt{k^2 - j_1^2 r^2}|\widetilde{m_{11}\mp 1}\rangle,$$

$$A_{\substack{01\\10}}|\tilde{m}\rangle = \frac{\sqrt{k^2 - j_1^2 r^2}}{2r}\left\{\left(s - m_{11} - m_{23} \mp \frac{1}{2}\right)|\tilde{m}\rangle\right.$$

$$\left. + j_1 r^2 \frac{q-s\pm 1}{k^2 - j_1^2 r^2}|\tilde{m}\rangle - r(|\tilde{m}\rangle'_{12} - |\tilde{m}\rangle'_{12})\right\}. \tag{4.49}$$

The Hermiticity relation for operators A_{02}, A_{20} gives $k^2 - j_1^2 r^2 \geq 0$, which for $j_1 = 1$ imposes the restriction $k \geq r$. The action of the operators on the derived patterns can be found, using (4.44) by application of the operators to both sides of the equation $|m\rangle'_{12} = (|m_{12} + \iota_2\rangle - |m_{12} - \iota_2\rangle)/2\iota_2$. The eigenvalues of Casimir operators for the representation (4.49) can be obtained by substituting the components (4.46) in (4.36). They are as follows:

$$C_1(j_1, \iota_2) = q + m_{23}, \quad C_2(j_1, \iota_2) = 2k^2, \quad C_3(j_1, \iota_2) = 3k^2(q+1). \tag{4.50}$$

They are all different from zero and independent, as it must be for non-degenerate irreducible representations of algebra $u(3; j_1, \iota_2)$. We also notice that the spectrum (4.50) coincides with the spectrum (4.41) of Casimir operators for algebra $u(3; \iota_1, j_2)$.

For the sake of convenient applications (and interpretation) we have fixed the indices of generators A_{pr}; for this reason $u(3; 1, \iota_2)$ and $u(3; \iota_1, j_2)$ turned out to be different algebras in our case. Rejecting this agreement, it is easy to prove that these algebras are isomorphic. The representation (4.39),

(4.40) is realized in discrete basis, generated by the chain of subalgebras $u(3; \iota_1, 1) \supset u(2; 1) \supset u(1)$ and described by the patterns:

$$\left| \begin{array}{ccc} k & m_{23} & -k \\ & m_{12} & m_{22} \\ & m_{11} & \end{array} \right\rangle, \qquad \begin{array}{c} m_{23} \in \mathbb{Z}, \ k \geq 0, \\ m_{12} \geq m_{23} \geq m_{22}, \ m_{12} \geq m_{11} \geq m_{22}, \\ m_{12}, m_{22}, m_{11} \in \mathbb{Z}, \end{array} \tag{4.51}$$

whereas the representation (4.49) is realized in continuous basis, generated by the expansion $u(3; 1, \iota_2) \supset u(2; \iota_2) \supset u(1)$ and described by the patterns

$$\left| \begin{array}{ccc} k & m_{23} & -k \\ & r & -r \\ & m_{11} & \end{array} \right\rangle, \qquad \begin{array}{c} m_{23}, m_{11} \in \mathbb{Z}, \\ k \geq r \geq 0, \end{array} \tag{4.52}$$

where besides A_{00}, A_{11}, A_{22}, the operator $A_{01} + A_{10}$ is also diagonal in this basis.

Thus, contractions over different parameters, leading to isomorphic algebras, give the description of the same representation of contracted algebra in different bases, generated by canonical chains of subalgebras.

4.2.4   *Two-dimensional contraction*

The structure of the contracted algebra $u(3; \iota)$ is as follows: $u(3; \iota) = T_6 \uplus (\{A_{00}\} \oplus \{A_{11}\} \oplus \{A_{22}\})$, where nilpotent subalgebra T_6 is spanned over generators A_{pr}, $p, r = 1, 2$. The explicit form of the generators of irreducible representations of the algebra can be obtained by putting either $j_1 = \iota_1$, $j_2 = \iota_2$ in (4.33) or $j_2 = \iota_2$ in (4.39), or from (4.45) for $j_1 = \iota_1$. All three approaches lead to the same result:

$$A_{00}|m\rangle = \left(m_{23} + \frac{m_{13} + m_{33}}{\iota_1 \iota_2} - \frac{m_{12} + m_{22}}{\iota_2} \right) |m\rangle,$$

$$A_{11}|m\rangle = \left(\frac{m_{12} + m_{22}}{\iota_2} - m_{11} \right) |m\rangle, \qquad A_{22}|m\rangle = m_{11}|m\rangle,$$

$$A_{02}|m\rangle = \frac{2ab}{m_{12} - m_{22}} |m_{11} - 1\rangle, \qquad A_{12}|m\rangle = a|m_{11} - 1\rangle,$$

$$A_{20}|m\rangle = \frac{2ab}{m_{12} - m_{22}} |m_{11} + 1\rangle, \qquad A_{21}|m\rangle = a|m_{11} + 1\rangle,$$

$$A_{01}|m\rangle = \frac{b}{m_{12} - m_{22}} \left\{ \frac{m_{12} + m_{22}}{\iota_2}|m\rangle - \left[m_{11} + m_{23} + \right. \right.$$

$$\left. \left. + \frac{3m_{12} + m_{22}}{2(m_{12} - m_{22})} \right] |m\rangle - m_{12}|m\rangle'_{12} - m_{22}|m\rangle'_{22} \right\},$$

$$A_{10}|m\rangle = \frac{b}{m_{12} - m_{22}} \left\{ \frac{m_{12} + m_{22}}{\iota_2}|m\rangle - \left[m_{11} + m_{23} + \right. \right.$$

$$\left. \left. + \frac{m_{12} + 3m_{22}}{2(m_{12} - m_{22})} \right] |m\rangle + m_{12}|m\rangle'_{12} + m_{22}|m\rangle'_{22} \right\},$$

$$a = \frac{1}{-m_{12}m_{22}}, \quad b = \frac{1}{-m_{13}m_{33}}. \tag{4.53}$$

The requirement of determinacy of the operators A_{00}, A_{11}, and the Hermiticity condition give for the components of the pattern $|m\rangle$:

$$m_{13} = k + \iota_1\iota_2\frac{q}{2}, \quad m_{33} = -k + \iota_1\iota_2\frac{q}{2}, \quad k \geq 0,\ q \in \mathbb{R},$$

$$m_{12} = r + \iota_2\frac{s}{2}, \quad m_{22} = -r + \iota_2\frac{s}{2}, \quad r \geq 0,\ s \in \mathbb{R}. \tag{4.54}$$

The substitution of these expressions in (4.53) leads to the representation operators

$$A_{00}|\tilde{m}\rangle = (m_{23} + q - s)|\tilde{m}\rangle, \quad A_{11}|\tilde{m}\rangle = (s - m_{11})|\tilde{m}\rangle,$$

$$A_{22}|\tilde{m}\rangle = m_{11}|\tilde{m}\rangle, \quad A_{12}|\tilde{m}\rangle = r\widetilde{|m_{11} - 1\rangle}, \quad A_{21}|\tilde{m}\rangle = r\widetilde{|m_{11} + 1\rangle},$$

$$A_{02}|\tilde{m}\rangle = k\widetilde{|m_{11} - 1\rangle}, \quad A_{20}|\tilde{m}\rangle = k\widetilde{|m_{11} + 1\rangle},$$

$$A_{01}|\tilde{m}\rangle = \frac{k}{2r}\left(s - m_{11} - m_{23} - \frac{1}{2}\right)|\tilde{m}\rangle - \frac{k}{2}(|\tilde{m}\rangle'_{12} - |\tilde{m}\rangle'_{22}),$$

$$A_{10}|\tilde{m}\rangle = \frac{k}{2r}\left(s - m_{11} - m_{23} + \frac{1}{2}\right)|\tilde{m}\rangle + \frac{k}{2}(|\tilde{m}\rangle'_{12} - |\tilde{m}\rangle'_{22}), \tag{4.55}$$

where $|\tilde{m}\rangle$ means the pattern

$$|\tilde{m}\rangle = \left| \begin{matrix} k & & m_{23} & & -k \\ & r & & -r & \\ & & m_{11} & & \end{matrix} \right\rangle, \quad \begin{matrix} m_{11},\ m_{23} \in \mathbb{Z}, \\[6pt] k \geq r \geq 0. \end{matrix} \tag{4.56}$$

It is worth mentioning that the operator $A_{01} + A_{10}$ is diagonal in basis $|\tilde{m}\rangle$, and the spectrum of Casimir operators for representations (4.55) of algebra

$u(3;\iota)$ is given by the same formulas (4.41), (4.50), as in the case of algebras $u(3;\iota_1,j_2)$, $u(3;j_1,\iota_2)$.

4.3 Representations of unitary algebras $u(n;j)$

4.3.1 *Operators of representation*

The standard Gel'fand–Tsetlin notations [Barut and Raczka (1977)] correspond to diminishing chain of subalgebras $u(n) \supset u(n-1) \supset \ldots \supset u(2) \supset u(1)$, where $u(n) = \{E_{kr}, k, r = 1, 2, \ldots, n\}, \ldots, u(2) = \{E_{kr}, k, r, = 1, 2\}$, $u(1) = \{E_{11}\}$. We shall now use another imbedding of subalgebra into algebra, which leads to the chain of subalgebras $u(n;j_1,j_2,\ldots,j_{n-1}) \supset u(n-1;j_2,\ldots,j_{n-1}) \supset \ldots \supset u(2;j_{n-1}) \supset u(1)$, where $u(n;j_1,j_2,\ldots,j_{n-1}) = \{A_{sp}, s,p = 0, 1, \ldots n-1\}, u(n-1;j_2,\ldots,j_{n-1}) = \{A_{sp}, s,p = 1, 2, \ldots, n-1\}, \ldots, u(2;j_{n-1}) = \{A_{sp}, s,p = n-2, n-1\}$, $u(1) = \{A_{n-1,n-1}\}$. To pass from standard notations to ours, it is necessary to change index k of the generator for index $n-k$ and to leave unchanged the numbering of components in Gel'fand–Tsetlin patterns.

To determine the representations of algebra $u(n)$ it is sufficient to give the action of generators E_{kk}, $E_{k,k+1}$, $E_{k+1,k}$, and to find the rest of generators from commutators. In our notations it is sufficient to know generators $A_{n-k,n-k}$, $A_{n-k,n-k-1}$, $A_{n-k-1,n-k}$, which are transformed under transition from $u(n)$ to $u(n;j)$ as follows:

$$A_{n-k,n-k-1} = j_{n-k}A^*_{n-k,n-k-1}(\rightarrow),$$
$$A_{n-k-1,n-k} = j_{n-k}A^*_{n-k-1,n-k}(\rightarrow), \quad k = 1, 2, \ldots, n, \qquad (4.57)$$
$$A_{n-k,n-k} = A^*_{n-k,n-k}(\rightarrow),$$

where j_{n-k} for nilpotent value plays the role of a tending-to-zero parameter in Wigner-Inönü contraction, and $A^*(\rightarrow)$ is a singularly transformed generator. Giving a singular transformation is equivalent to giving the transformation rule for components of Gel'fand–Tsetlin pattern

$$|m\rangle = \begin{vmatrix} m^*_{1n} & & m^*_{2n} & \cdots & m^*_{n-1,n} & & m^*_{nn} \\ & m^*_{1,n-1} & & m^*_{2,n-1} & \cdots & m^*_{n-1,n-1} & \\ & & & \cdots\cdots\cdots & & & \\ & & & m^*_{12} & m^*_{22} & & \\ & & & m^*_{11} & & & \end{vmatrix}, \qquad (4.58)$$

$$m^*_{pk} \geq m^*_{p,k-1} \geq m^*_{p+1,k}, \quad k = 2, 3, \ldots, n, \ p = 1, 2, \ldots, n-1,$$
$$m^*_{1n} \geq m^*_{2n} \geq \ldots \geq m^*_{nn}$$

under the transition from $u(n)$ to $u(n; j)$. Defining this transformation by

$$m_{1k} = m_{1k}^* J_k, \quad m_{kk} = m_{kk}^* J_k, \quad J_k = \prod_{l=n-k+1}^{n-1} j_l,$$

$$m_{pk} = m_{pk}^*, \quad p = 2, 3, \ldots, k-1, \; k = 2, 3, \ldots, n,$$

(4.59)

we obtain the pattern $|m\rangle$, of which the components m_{pk} are integers, and the components m_{1k}, m_{kk} can be complex or nilpotent numbers. They satisfy the inequalities

$$m_{pk} \geq m_{p,k-1} \geq m_{p+1,k}, \; k = 2, 3, \ldots, n, \; p = 2, 3, \ldots, n-2$$

$$\frac{m_{1k}}{J_k} \geq \frac{m_{1,k-1}}{J_{k-1}} \geq m_{2k}, \quad m_{k-1,k} \geq \frac{m_{k-1,k-1}}{J_{k-1}} \geq \frac{m_{kk}}{J_k},$$

$$\frac{m_{1n}}{J_n} \geq m_{2n} \geq m_{3n} \geq \ldots \geq m_{n-1,n} \geq \frac{m_{nn}}{J_n},$$

which are interpreted for dual and imaginary values of parameters j according to the rules, described in section 4.2.1.

Substituting (4.59) in the known expressions for generators of algebra and taking into account (4.57), we find the operators of representation of algebra $u(n; j)$

$$A_{n-k,n-k}|m\rangle = \left[\frac{m_{1k} + m_{kk}}{J_k} - \frac{m_{1,k-1} + m_{k-1,k-1}}{J_{k-1}} \right.$$

$$\left. + m_{k-1,k} + \sum_{s=2}^{k-2}(m_{sk} - m_{s,k-1}) \right] |m\rangle, \quad k = 1, 2, \ldots, n,$$

$$A_{n-k-1,n-k}|m\rangle = \frac{1}{J_k} \left[a_k^1(m)|m_{1k} - J_k\rangle + a_k^k(m)|m_{kk} - J_k\rangle \right]$$

$$+ j_{n-k+1} \sum_{s=2}^{k-1} a_k^s(m)|m_{sk} - 1\rangle,$$

$$A_{n-k,n-k-1}|m\rangle = \frac{1}{J_k} \left[b_k^1(m)|m_{1k} + J_k\rangle + b_k^k(m)|m_{kk} + J_k\rangle \right]$$

$$+ j_{n-k+1} \sum_{s=2}^{k-1} b_k^s(m)|m_{sk} + 1\rangle, \quad k = 1, 2, \ldots, n-1,$$

(4.60)

where

$$a_k^1(m) = \left\{ \frac{\prod_{p=2}^{k}(J_k l_{p,k+1} - l_{1k} + J_k)\prod_{p=2}^{k-2}(J_k l_{p,k-1} - l_{1k})}{\prod_{p=2}^{k-1}(J_k l_{pk} - l_{1k} + J_k)(J_k l_{pk} - l_{1k})} \right\}^{1/2}$$

$$\times \left\{ -\frac{(l_{1,k+1} - j_{n-k}l_{1k} + J_{k+1})(l_{k+1,k+1} - j_{n-k}l_{1k} + J_{k+1})}{(l_{kk} - l_{1k} + J_k)} \right.$$

$$\times \left. \frac{(l_{1,k-1}j_{n-k+1} - l_{1k})(l_{k-1,k-1}j_{n-k+1} - l_{1k})}{(l_{kk} - l_{1k})} \right\}^{1/2},$$

$$a_k^s(m) = \left\{ \frac{\prod_{p=2}^{k}(l_{p,k+1} - l_{sk} + 1)\prod_{p=2}^{k-2}(l_{p,k-1} - l_{sk})}{\prod_{p=2,p\neq s}^{k-1}(l_{pk} - l_{sk} + 1)(l_{pk} - l_{sk})} \right\}^{1/2}$$

$$\times \left\{ -\frac{(l_{1,k+1} - J_{k+1}l_{sk} + J_{k+1})(l_{k+1,k+1} - J_{k+1}l_{sk} + J_{k+1})}{(l_{1k} - J_k l_{sk} + J_k)(l_{1k} - J_k l_{sk})} \right.$$

$$\times \left. \frac{(l_{1,k-1} - J_{k-1}l_{sk})(l_{k-1,k-1} - J_{k-1}l_{sk})}{(l_{kk} - J_k l_{sk} + J_k)(l_{kk} - J_k l_{sk})} \right\}^{1/2}, \quad 1 < s < k,$$

$$b_k^1(m) = \left\{ \frac{\prod_{p=2}^{k}(J_k l_{p,k+1} - l_{1k})\prod_{p=2}^{k-2}(J_k l_{p,k-1} - l_{1k} - J_k)}{\prod_{p=2}^{k-1}(J_k l_{pk} - l_{1k})(J_k l_{pk} - l_{1k} - J_k)} \right\}^{1/2}$$

$$\times \left\{ -\frac{(l_{1,k+1} - j_{n-k}l_{1k})(l_{k+1,k+1} - j_{n-k}l_{1k})}{(l_{kk} - l_{1k})} \right.$$

$$\times \left. \frac{(l_{1,k-1}j_{n-k+1} - l_{1k} - J_k)(l_{k-1,k-1}j_{n-k+1} - l_{1k} - J_k)}{(l_{kk} - l_{1k} - J_k)} \right\}^{1/2},$$

$$b_k^s(m) = \left\{ \frac{\prod_{p=2}^{k}(l_{p,k+1} - l_{sk})\prod_{p=2}^{k-2}(l_{p,k-1} - l_{sk} - 1)}{\prod_{p=2,p\neq s}^{k-1}(l_{pk} - l_{sk})(l_{pk} - l_{sk} - 1)} \right\}^{1/2}$$

$$\times \left\{ -\frac{(l_{1,k+1} - J_{k+1}l_{sk})(l_{k+1,k+1} - J_{k+1}l_{sk})}{(l_{1k} - J_k l_{sk})(l_{1k} - J_k l_{sk} - J_k)} \right.$$

$$\times \left. \frac{(l_{1,k-1} - J_{k-1}l_{sk} - J_{k-1})(l_{k-1,k-1} - J_{k-1}l_{sk} - J_{k-1})}{(l_{kk} - J_k l_{sk})(l_{kk} - J_k l_{sk} - J_k)} \right\}^{1/2},$$

$$1 < s < k. \tag{4.61}$$

The expression for $a_k^k(m)$ can be derived from $a_k^1(m)$ by changing l_{1k} for l_{kk} and l_{kk} for l_{1k}. The same substitution turns $b_k^1(m)$ into $b_k^k(m)$. Components m are related with components l by the equations

$$l_{1k} = m_{1k} - J_k, \, l_{kk} = m_{kk} - kJ_k, \, l_{sk} = m_{sk} - s, 1 < s < k. \tag{4.62}$$

As it can be shown by direct checking, the operators (4.60) satisfy the commutation relations (4.34) of algebra $u(n; j)$. Therefore, they give a representation of algebra. Considering the action of the raising operators $A_{n-k,n-k-1}$ on the "highest weight" vector z_h, described by the pattern $|m\rangle$ for the maximal values of components, and the action of the lowering operators $A_{n-k-1,n-k}$ on the "lowest weight" vector z_l, described by the pattern $|m\rangle$ for the minimal values of components as in section 4.2.1, we find that for nilpotent or imaginary values of all or some parameters j the space of representation is infinite-dimensional and does not contain subspaces invariant with respect to the operators (4.60), because in taking any basis vector and acting on it with operators A_{kr} by the required number of times, we obtain all basis vectors in the space of representation. Therefore, the representation (4.60) is irreducible.

Though the initial representation of algebra $u(n)$ is Hermitian, the irreducible representation (4.60) of algebra $u(n; j)$, in general, is not Hermitian. Therefore, if we want the representation (4.60) to be Hermitian, it is necessary to fulfill the relations $A_{pp}^\dagger = A_{pp}$, $p = 0, 1, \ldots, n-1$, $A_{kp} = A_{pk}^\dagger$, which for matrix elements of operators can be written as follows:

$$\langle m|A_{pp}|m\rangle = \overline{\langle m|A_{pp}|m\rangle}, \quad \langle n|A_{kp}|m\rangle = \overline{\langle m|A_{pk}|n\rangle}, \tag{4.63}$$

where the bar means complex conjugation.

4.3.2 Spectrum of Casimir operators

The components m_{kn}^* of the upper row of the pattern (4.58) (components of the highest weight) completely determine the irreducible representation of algebra $u(n)$. Perelomov and Popov (1966a), Leznov, Malkin and Man'ko (1977) found the eigenvalues of Casimir operators, expressing them in terms of components of the highest weight. For unitary algebra $u(n)$ the spectrum of Casimir operators can be written as follows:

$$C_q^*(m^*) = \mathrm{Tr}\,(a^{*q}E), \tag{4.64}$$

where E is the matrix of dimension n, all elements of which are equal to a unit, and matrix a^* is as follows

$$a_{ps}^* = (m_{pn}^* + n - p)\delta_{ps} - w_{sp}, \quad s, p = 1, 2, \ldots, n. \tag{4.65}$$

Here $w_{sp} = 1$ for $s < p$ $w_{sp} = 0$ for $s > p$.

Under transition from algebra $u(n)$ to algebra $u(n; j)$, $j = (j_1, \ldots, j_{n-1})$ the components of the highest weight are transformed according to (4.59):

$m_{1n} = J m^*_{1n}$, $m_{nn} = J m^*_{nn}$, $m_{sn} = m^*_{sn}$, $s = 2, 3, \ldots, n-1$, $J = \prod_{l=1}^{n-1} j_l$.
Let us define the matrix $a(j)$ as

$$a(j) = J a^*(\rightarrow), \tag{4.66}$$

where $a^*(\rightarrow)$ is the matrix (4.65), in which the components m^*_{pn} are substituted by their expressions in terms of m_{pn}, i.e. $a_{11}(\rightarrow) = n - 1 + m_{1n} J^{-1}$, $a_{nn}(\rightarrow) = m_{nn} J^{-1}$, and the remaining matrix elements are given by (4.65). Then matrix $a(j)$ is as follows:

$$a_{11}(j) = m_{1n} + J(n-1), \quad a_{nn}(j) = m_{nn},$$
$$a_{ps}(j) = J[(m_{pn} + n - p)\delta_{ps} - w_{sp}], \ p, s = 2, 3, \ldots, n-2. \tag{4.67}$$

Casimir operators are transformed according to (1.98), i.e. $C_{2k}(j) = J^{2k} C^*_{2k}(\rightarrow)$, $C_{2k+1}(j) = J^{2k} C^*_{2k+1}(\rightarrow)$. Their spectra are transformed in just the same way. Therefore, the spectrum of Casimir operators for algebra $u(n; j)$ is as follows:

$$C_{2k}(m) = J^{2k} \mathrm{Tr}\{a^{*2k}(\rightarrow)E\} = \mathrm{Tr}\{[J a^*(\rightarrow)]^{2k} E\} = \mathrm{Tr}\{a^{2k}(j)E\},$$
$$C_{2k+1}(m) = J^{2k} \mathrm{Tr}\{a^{*2k+1}(\rightarrow)E\} = \mathrm{Tr}\{a^*(\rightarrow)[J a^*(\rightarrow)]^{2k} E\}$$
$$= \mathrm{Tr}\{a^*(\rightarrow)a^{2k}(j)E\} = \frac{1}{J} \mathrm{Tr}\{a^{2k+1}(j)E\}, \tag{4.68}$$

where $2k$ and $2k+1$ take all integer values from 1 to n. In particular,

$$C_1(m) = \frac{m_{1n} + m_{nn}}{J} + \sum_{s=2}^{n-1} m_{sn},$$

$$C_2(m) = m_{1n}^2 + m_{nn}^2 + J(n-1)(m_{1n} - m_{nn})$$
$$+ J^2 \sum_{s=2}^{n-1} m_{sn}(m_{sn} + n + 1 - 2s) \tag{4.69}$$

are the eigenvalues of the first two Casimir operators of algebra $u(n; j)$ on the irreducible representation.

4.3.3   *Possible variants of contractions of irreducible representations*

This section will discuss, in brevity, on the contractions of irreducible representations, however, keeping in mind that the corresponding considerations are valid for imaginary values of parameters j as well. The transformations (4.59) of components in Gel'fand–Tsetlin pattern have been chosen in such a way that the eigenvalues of even order Casimir operators would differ from zero under contractions. However, the variant (4.59) (we call it basic) is not unique. Easily seen from (4.68), (4.69), the same goal can be achieved, transforming any two components of the upper row according to the rule $m = Jm^*$ and leaving the other components of this row unchanged.

In this case, what happens with the initial irreducible representation, if we consider the contraction $j_1 = \iota_1$, $j_2 = \cdots = j_{n-1} = 1$? The transformation rule for generators remains unchanged: $A = (\prod_k j_k)A^*(\rightarrow)$. Only the expressions $A^*(\rightarrow)$ for singularly transformed operators of representation are modified as well as the inequalities for components of Gel'fand–Tsetlin pattern in comparison with the same contraction $j_1 = \iota_1$ in the basic variant. The eigenvalues of Casimir operators do not depend on the components m_{1n}, m_{nn}, as in basic variant, but on the other two components of the upper row.

Thus, each of $\binom{n}{2} = n(n-1)/2$ variants of transition from the irreducible representation of algebra $u(n)$ gives under contractions its own irreducible representation of algebra $u(n; j)$, of which the spectrum of Casimir operators is determined by its own two components of the upper row of Gel'fand–Tsetlin pattern. In this case all variants of transition are of the general type, i.e. lead to nonzero spectrum of all Casimir operators, even when all parameters j take nilpotent values.

The considerations brought above are valid for each algebra $u(k; j')$, $k = 2, 3, \ldots, n-1$ in the chain of subalgebras, described in section 4.3.1, i.e. for each subalgebra there are $\binom{k}{2} = k(k-1)/2$ variants of transition from the irreducible representation of subalgebra $u(k)$ to the general irreducible representations of subalgebra $u(k; j')$. The latter determines Gel'fand–Tsetlin basis. Therefore each of $\binom{n}{2}$ variants of transition from the irreducible representation of algebra $u(n)$ to the irreducible representations of algebra $u(n; j)$ can be written in $N_{n-1} = \sum_{k=2}^{n-1} \binom{k}{2}$ different bases, corresponding to different variants of transformation of Gel'fand–Tsetlin pattern components in the rows with numbers $k = 2, 3, \ldots, n-1$. In the sections 4.3.1 and 4.3.2 we have described a basic variant, in which the first and the last components of the rows with numbers $k = 2, 3, \ldots, n$ undergo transformation. It is clear that, if necessary, similar relations can be written for each of $N_k = \sum_{k=2}^{n} \binom{k}{2}$ variants.

4.4 Representations of orthogonal algebras

4.4.1 *Algebra so(3; j)*

Despite the fact that algebra $so(3)$ is isomorphic to algebra $u(2)$, we shall consider it separately, because algebras $so(3; j)$, $j = (j_1, j_2)$, in contrast to $u(2; j_1)$, allow contraction over two parameters. Under transition from $so(3)$ to $so(3; j)$ generators are transformed as follows: $X_{01} = j_1 X_{01}^*(\rightarrow)$, $X_{02} = j_1 j_2 X_{02}^*(\rightarrow)$, $X_{12} = j_2 X_{12}^*(\rightarrow)$, and the only Casimir operator is transformed as $C_2(j) = j_1^2 j_2^2 C_2^*(\rightarrow)$. Gel'fand and Tsetlin (1950) have also found irreducible representations of algebra $so(3)$ in the basis, determined by the chain of subalgebras $so(3) \supset so(2)$; they are given by operators

$$X_{12}^*|m^*\rangle = i m_{11}^*|m^*\rangle,$$

$$X_{01}^*|m^*\rangle = \frac{1}{2}\left\{\sqrt{(m_{12}^* - m_{11}^*)(m_{12}^* + m_{11}^* + 1)}|m_{11}^* + 1\rangle\right.$$
$$\left. - \sqrt{(m_{12}^* + m_{11}^*)(m_{12}^* - m_{11}^* + 1)}|m_{11}^* - 1\rangle\right\},$$

$$X_{02}^*|m^*\rangle = \frac{i}{2}\left\{\sqrt{(m_{12}^* - m_{11}^*)(m_{12}^* + m_{11}^* + 1)}|m_{11}^* + 1\rangle\right.$$
$$\left. + \sqrt{(m_{12}^* + m_{11}^*)(m_{12}^* - m_{11}^* + 1)}|m_{11}^* - 1\rangle\right\}, \quad (4.70)$$

where the patterns $|m^*\rangle$, enumerating the elements of Gel'fand–Tsetlin orthonormal basis, are $|m^*\rangle = \left|\begin{matrix} m_{12}^* \\ m_{11}^* \end{matrix}\right\rangle$, $|m_{11}^*| \leq m_{12}^*$ and the components m_{11}^*, m_{12}^* are simultaneously either integer, or half-integer. The spectrum of Casimir operator is

$$C_2^*(m_{12}^*) = m_{12}^*(m_{12}^* + 1). \quad (4.71)$$

The component m_{11}^* is the eigenvalue of the operator X_{12}^*; for this reason its transformation is determined by the transformation X_{12}^*, i.e. $m_{11} = j_2 m_{11}^*$. The transformation of the component m_{12}^* can be found from the requirement of determinacy and the nonzero spectrum of Casimir operator

$$C_2(m_{12}) = j_1^2 j_2^2 C_2^*(\rightarrow)$$
$$= j_1^2 j_2^2 \frac{m_{12}}{j_1 j_2}\left[\frac{m_{12}}{j_1 j_2} + 1\right] = m_{12}(m_{12} + j_1 j_2), \quad (4.72)$$

under contractions, i.e. $m_{12} = j_1 j_2 m_{12}^*$. Then we find from (4.70) the operators of representation of algebra $so(3; j)$:

$$X_{12}|m\rangle = i m_{11}|m\rangle,$$

$$X_{01}|m\rangle = \frac{1}{2j_2}\Big\{\sqrt{(m_{12} - j_1 m_{11})(m_{12} + j_1 m_{11} + j_1 j_2)}|m_{11} + j_2\rangle$$

$$- \sqrt{(m_{12} + j_1 m_{11})(m_{12} - j_1 m_{11} + j_1 j_2)}|m_{11} - j_2\rangle\Big\},$$

$$X_{02}|m\rangle = \frac{i}{2}\Big\{\sqrt{(m_{12} - j_1 m_{11})(m_{12} + j_1 m_{11} + j_1 j_2)}|m_{11} + j_2\rangle$$

$$+ \sqrt{(m_{12} + j_1 m_{11})(m_{12} - j_1 m_{11} + j_1 j_2)}|m_{11} - j_2\rangle\Big\}, \qquad (4.73)$$

where the components of the pattern $|m\rangle$ satisfy the inequalities $|m_{11}| \leq m_{12}/j_1$. The operators (4.73) satisfy commutation relations of algebra $so(3; j)$, which can be checked directly. The representation is irreducible. This can be established, as in the case of unitary algebras, by the action of raising and lowering operators on the vectors of the highest and lowest weights.

For $j_1 = \iota_1$ the relations (4.73) give irreducible representation of inhomogeneous algebra $so(3; \iota_1, j_2) = io(2; j_2) = \{X_{01}, X_{02}\} \oplus \{X_{12}\}$:

$$X_{12}|m\rangle = i m_{11}|m\rangle,$$

$$X_{01}|m\rangle = \frac{1}{2j_2} m_{12}(|m_{11} + j_2\rangle - |m_{11} - j_2\rangle), \qquad (4.74)$$

$$X_{02}|m\rangle = \frac{i}{2} m_{12}(|m_{11} + j_2\rangle + |m_{11} - j_2\rangle),$$

where m_{11} is integer or half-integer, and $m_{12} \in \mathbb{R}$, $m_{12} \geq 0$. The eigenvalues of Casimir operator are $C_2(\iota_1, j_2) = m_{12}^2$.

From (4.73) we obtain the irreducible representation of algebra $so(3; j_1, \iota_2) = \{X_{02}, X_{12}\} \oplus \{X_{01}\}$ for $j_2 = \iota_2$:

$$X_{12}|m\rangle = i m_{11}|m\rangle, \qquad X_{02}|m\rangle = i\sqrt{m_{12}^2 - j_1^2 m_{11}^2}|m\rangle,$$

$$X_{01}|m\rangle = \sqrt{m_{12}^2 - j_1^2 m_{11}^2}|m\rangle_{11}' - \frac{j_1^2 m_{11}}{2\sqrt{m_{12}^2 - j_1^2 m_{11}^2}}|m\rangle, \qquad (4.75)$$

where m_{11}, $m_{12} \in \mathbb{R}$ and $|m_{11}| \leq m_{12}$. Refusing to fix the coordinates x_0, x_1, x_2 in $\mathbf{R}_3(j)$, where group $SO(3; j)$ acts, we can easily prove isomorphism of algebras $so(3; \iota_1, 1)$ and $so(2; 1, \iota_2)$. Then (4.74) for $j_2 = 1$

and (4.75) for $j_1 = 1$ give the description of irreducible representation of algebra $io(2)$ in discrete and continuous basis, correspondingly, in infinite-dimensional representation space. The discrete basis consists of the eigenvectors of compact operator X_{12} with integer or half-integer eigenvalues $-\infty < m_{11} < \infty$. The continuous basis consists of generalized eigenvectors of noncompact operator X_{12}, the eigenvalues of which are $m_{11} \in \mathbb{R}$, $|m_{11}| \le m_{12}$. A close approach to contraction and irreducible representations of orthogonal algebras $so(3)$, $so(5)$, related with singular transformation of the components of Gel'fand–Tsetlin patterns has been considered by Celeghini and Tarlini (1981a,b, 1982).

Two-dimensional contraction $j_1 = \iota_1$, $j_2 = \iota_2$ gives irreducible representation of (two-dimensional) Galilean algebra $so(3; \iota)$:

$$X_{01}|m\rangle = m_{12}|m\rangle'_{11}, \quad X_{12}|m\rangle = im_{11}|m\rangle, \quad X_{02}|m\rangle = im_{12}|m\rangle, \quad (4.76)$$

where $m_{12}, m_{11} \in \mathbb{R}$, $m_{12} \ge 0$, $-\infty < m_{11} < \infty$ and $C_2(\iota) = m_{12}^2$. The action of generators on the derivative $|m\rangle'_{11}$ leads to the result found by applying a generator to both sides of equation $|m\rangle'_{11} = (|m_{11} + \iota_2\rangle - |m_{11} - \iota_2\rangle)/2\iota_2$. In particular, $X_{12}|m\rangle'_{11} = im_{11}|m\rangle'_{11} + i|m\rangle$.

4.4.2   *Algebra so(4; j)*

To determine the irreducible representation of algebra $so(4; j)$, $j = (j_1, j_2, j_3)$, it is sufficient to give a representation of the generators X_{01}, X_{12}, X_{23}. Using the formulae from monograph [Barut and Raczka (1977)], we can change the indices of generators according to the rule: $4 \to 0$, $3 \to 1$, $2 \to 2$, $1 \to 3$. Then Gel'fand–Tsetlin representation corresponds to the chain of subalgebras $so(4; \mathbf{j}) \supset so(3; j_2, j_3) \supset so(2; j_3)$, where $so(4; j) = \{X_{rs}, r < s, r, s = 0, 1, 2, 3\}$, $so(3; j_2, j_3) = \{X_{rs}, r < s, r, s = 1, 2, 3\}$, $so(2; j_3) = \{X_{23}\}$. The representation of generators X_{23}, X_{12} is given by (4.73), where the indices of the generators and parameters j must be increased by 1, and pattern $|m\rangle = \left|\begin{matrix} m_{12} \\ m_{11} \end{matrix}\right\rangle$ has to be substituted for

$$|m\rangle = \left|\begin{matrix} m_{13} & & m_{23} \\ & m_{12} & \\ & m_{11} & \end{matrix}\right\rangle. \quad (4.77)$$

The rule of transformation for the components m_{12}^*, m_{11}^* can be found by consideration of algebra $so(3; j)$: $m_{11} = j_3 m_{11}^*$, $m_{12} = j_2 j_3 m_{23}^*$, $|m_{11}| \le m_{12}/j_2$. It remains to derive the transformation of the components

m_{13}^*, m_{23}^*, which determine the irreducible representation. For this we consider the spectrum of Casimir operators for algebra $so(4)$, found by Leznov, Malkin and Man'ko (1977); Perelomov and Popov (1966b),

$$C_2^* = m_{13}^*(m_{13}^* + 2) + m_{23}^{*2}, \quad C_2^{*\prime} = -(m_{13}^* + 1)m_{23}^*, \tag{4.78}$$

as well as the rule of transformation for Casimir operators under transition from $so(4)$ to $so(4;j)$

$$C_2(j) = j_1^2 j_2^2 j_3^2 C_2^*(\rightarrow), \quad C_2'(j) = j_1 j_2^2 j_3 C_2^{*\prime}(\rightarrow). \tag{4.79}$$

Requiring the eigenvalues of operators $C_2(j)$ and $C_2'(j)$ to be determinate expressions under contractions, we get from (4.78), (4.79) for $C_2'(j)$

$$m_{13} m_{23} = j_1 j_2^2 j_3 m_{13}^* m_{23}^*. \tag{4.80}$$

This equation (if transformations of the components m_{13}, m_{23} involve only the first powers of parameters j) gives possible rules of transformation of these components.

Let us write down the possible variants of transformations of irreducible representations of algebra $so(4)$ into representations of algebra $so\,(4;j)$ as well as the transformed spectra of Casimir operators

(1) $m_{13} = j_1 j_2 m_{13}^*$, $\quad m_{23} = j_2 j_3 m_{23}^*$,

$$C_2(j) = j_3^2 m_{13}(m_{13} + 2j_1 j_2) + j_1^2 m_{23}^2,$$
$$C_2'(j) = -(m_{13} + j_1 j_2)m_{23}. \tag{4.81}$$

(2) $m_{13} = j_2 m_{13}^*$, $\quad m_{23} = j_1 j_2 j_3 m_{23}^*$,

$$C_2(j) = m_{23}^2 + j_1^2 j_3^2 m_{13}(m_{13} + 2j_2),$$
$$C_2'(j) = -(m_{13} + j_2)m_{23}. \tag{4.82}$$

(3) $m_{13} = j_1 j_2 j_3 m_{13}^*$,

$$C_2(j) = m_{13}(m_{13} + 2j_1 j_2 j_3) + j_1^2 j_3^2 m_{23}^2,$$
$$C_2'(j) = -(m_{13} + j_1 j_2 j_3)m_{23}. \tag{4.83}$$

Considering (4.78), (4.79), the following variants for only the operator $C_2(j)$ are admissible:

(4) $m_{13} = j_1 j_2 j_3 m_{13}^*$, $\quad m_{23} = m_{23}^*$,

$$C_2(j) = m_{13}(m_{13} + 2j_1 j_2 j_3) + j_1^2 j_2^2 j_3^2 m_{23}^2,$$
$$C_2'(j) = -j_2(m_{13} + j_1 j_2 j_3)m_{23}. \tag{4.84}$$

(5) $m_{13} = m_{13}^*, \quad m_{23} = j_1 j_2 j_3 m_{23}^*,$

$$C_2(j) = m_{23}^2 + j_1^2 j_2^2 j_3^2 m_{13}(m_{13} + 2),$$
$$C_2'(j) = -j_2(m_{13} + 1)m_{23},$$

(4.85)

and other variants of transformations of the components m_{13}^*, m_{23}^*, which do not include all the parameters j, up to variant $m_{13} = m_{13}^*$, $m_{23} = m_{23}^*$.

Considering the contraction for all the parameters $j = \iota$, we see that general nondegenerate (with nonzero eigenvalues of both Casimir operators) representations of the contracted algebra $so(4; \iota)$ come out only in the case of transformations (2) and (3). In the case of transformations (1) we get $C_2(\iota) = 0$, in the case of transformations (4), (5) — $C_2'(\iota) = 0$. Under other transformations both Casimir operators have zero spectrum.

Let us consider variant (3). In this case the components of the pattern (4.77) satisfy inequalities

$$\frac{m_{13}}{j_1 j_3} \geq |m_{23}|, \quad \frac{m_{13}}{j_1 j_3} \geq \frac{m_{12}}{j_3} \geq |m_{23}|, \quad \frac{m_{12}}{j_2} \geq |m_{11}|, \qquad (4.86)$$

interpreted for imaginary and nilpotent values of the parameters j according to the rules represented in 4.2.1. Using (4.83) and the rules of transformations for the generators $X_{01} = j_1 X_{01}^*$, $X_{02} = j_1 j_2 X_{02}^*$, $X_{03} = j_1 j_2 X_{03}^*$, we find the operators of irreducible representations of algebra $so(4; j)$:

$$X_{01}|m\rangle = im_{11}b|m\rangle - \frac{a(m_{12})}{j_2 j_3}\sqrt{m_{12}^2 - j_2^2 m_{11}^2}|m_{12} - j_2 j_3\rangle$$

$$+ \frac{a(m_{12} + j_2 j_3)}{j_2 j_3}\sqrt{(m_{12} + j_2 j_3)^2 - j_2^2 m_{11}^2}|m_{12} + j_2 j_3\rangle,$$

$$X_{02}|m\rangle = \frac{ib}{2}\left\{\sqrt{(m_{12} - j_2)(m_{12} + j_2 m_{11} + j_2 j_3)}|m_{11} + j_3\rangle\right.$$

$$\left. + \sqrt{(m_{12} + j_2 m_{11})(m_{12} - j_2 m_{11} + j_2 j_3)}|m_{11} - j_3\rangle\right\}$$

$$- \frac{a(m_{12})}{2j_3}\left\{\sqrt{(m_{12} - j_2 m_{11})(m_{12} - j_2 m_{11} - j_2 j_3)}\left|\begin{matrix}m_{12} - j_2 j_3\\ m_{11} + j_2\end{matrix}\right\rangle\right.$$

$$\left. - \sqrt{(m_{12} + j_2 m_{11})(m_{12} + j_2 m_{11} - j_2 j_3)}\left|\begin{matrix}m_{12} - j_2 j_3\\ m_{11} - j_3\end{matrix}\right\rangle\right\}$$

$$- \frac{a(m_{12} + j_2 j_3)}{2 j_3}$$

$$\times \left\{ \sqrt{(m_{12} + j_2 m_{11} + j_2 j_3)(m_{12} + j_2 m_{11} + 2 j_2 j_3)} \left| \begin{array}{c} m_{12} + j_2 j_3 \\ m_{11} + j_3 \end{array} \right\rangle \right.$$

$$\left. - \sqrt{(m_{12} - j_2 m_{11} + j_2 j_3)(m_{12} - j_2 m_{11} + 2 j_2 j_3)} \left| \begin{array}{c} m_{12} + j_2 j_3 \\ m_{11} - j_3 \end{array} \right\rangle \right\},$$

$$X_{03}|m\rangle = \frac{j_3 b}{2} \left\{ \sqrt{(m_{12} - j_2 m_{11})(m_{12} + j_2 m_{11} + j_2 j_3)} |m_{11} + j_3\rangle \right.$$

$$\left. - \sqrt{(m_{12} + j_2 m_{11})(m_{12} - j_2 m_{11} + j_2 j_3)} |m_{11} - j_3\rangle \right\}$$

$$+ \frac{ia(m_{12})}{2} \left\{ \sqrt{(m_{12} - j_2 m_{11})(m_{12} - j_2 m_{11} - j_2 j_3)} \left| \begin{array}{c} m_{12} - j_2 j_3 \\ m_{11} + j_3 \end{array} \right\rangle \right.$$

$$\left. + \sqrt{(m_{12} + j_2 m_{11})(m_{12} + j_2 m_{11} - j_2 j_3)} \left| \begin{array}{c} m_{12} - j_2 j_3 \\ m_{11} - j_3 \end{array} \right\rangle \right\}$$

$$+ \frac{i}{2} a(m_{12} + j_2 j_3)$$

$$\times \left\{ \sqrt{(m_{12} + j_2 m_{11} + j_2 j_3)(m_{12} + j_2 m_{11} + 2 j_2 j_3)} \left| \begin{array}{c} m_{12} + j_2 j_3 \\ m_{11} + j_3 \end{array} \right\rangle \right.$$

$$\left. + \sqrt{(m_{12} - j_2 m_{11} + j_2 j_3)(m_{12} - j_2 m_{11} + 2 j_2 j_3)} \left| \begin{array}{c} m_{12} + j_2 j_3 \\ m_{11} - j_3 \end{array} \right\rangle \right\},$$

$$a(m_{12}) = \sqrt{\frac{[(m_{13} + j_1 j_2 j_3)^2 - j_1^2 m_{12}^2](m_{12}^2 - j_3^2 m_{23}^2)}{m_{12}^2 (4 m_{12}^2 - j_2^2 j_3^2)}},$$

$$b = \frac{(m_{13} + j_1 j_2 j_3) m_{23}}{m_{12}(m_{12} + j_2 j_3)}. \tag{4.87}$$

Here we have presented all the generators of algebra $so(4; j)$ though it is noticeably sufficient to give only X_{01}.

The initial finite-dimensional irreducible representation of algebra $so(4)$ is Hermitian. The representation (4.87) of algebra $so(4; j)$ is irreducible but, in general, non-Hermitian. To obtain Hermitian representation, it is necessary to impose the requirement of Hermiticity: $X_{rs}^{\dagger} = -X_{rs}$ on the operators (4.87). It is difficult to find the restrictions implied by this requirement on the components of Gel'fand–Tsetlin patterns. Therefore the requirement of Hermiticity has to be checked in any particular case for the concrete values of the parameters j.

Considering variant (2), determined by (4.82), it turns out that the components of the pattern (4.77) satisfy inequalities

$$m_{13} \geq \frac{m_{23}}{j_1 j_3}, \quad m_{13} \geq \frac{m_{12}}{j_3} \geq \frac{|m_{23}|}{j_1 j_3}, \quad \frac{m_{12}}{j_2} \geq |m_{11}|, \qquad (4.88)$$

the operators X_{12}, X_{13}, X_{23} are described by (4.73), where the indices of parameters and generators have to be increased by one, and operators X_{0k}, $k = 1, 2, 3$ are given by (4.87), where functions b and $a(m_{12})$ are substituted by functions $\tilde{b}$, $\tilde{a}(m_{12})$ which are as follows:

$$\tilde{a}(m_{12}) = \sqrt{\frac{[j_3^2(m_{13} + j_2)^2 - m_{12}^2](j_1^2 m_{12}^2 - m_{23}^2)}{m_{12}(4m_{12}^2 - j_2^2 j_3^2)}},$$

$$\tilde{b} = \frac{(m_{13} + j_2)m_{23}}{m_{12}(m_{12} + j_2 j_3)}. \qquad (4.89)$$

4.4.3 *Contractions of representations of algebra so(4; j)*

Let us consider representations of algebra $so(4; \iota_1, j_2, j_3) = \{X_{0k}\} \supseteq so(3; j_2, j_3)$. Let the components m_{13}, m_{23} be transformed according to (4.83), i.e. $k = m_{23} = \iota_1 j_2 j_3 m_{13}^*$, $m_{23} = j_2 m_{23}^*$. The operators of the representation are described by (4.87), where

$$a(m_{12}) = k\sqrt{\frac{m_{12}^2 - j_3^2 m_{23}^2}{m_{12}^2(4m_{12}^2 - j_2^2 j_3^2)}}, \quad b = \frac{k m_{23}}{m_{12}(m_{12} + j_2 j_3)}. \qquad (4.90)$$

From the inequalities (4.86) for $j_2 = j_3 = 1$ we find $0 \leq |m_{23}| < \infty$, $m_{12} \geq |m_{23}|$, $|m_{11}| \leq m_{12}$, where $m_{11}, m_{12}, m_{23} \in \mathbb{Z}$, $k \in \mathbb{R}$ (the latter — from the requirement of Hermiticity for X_{01}). The spectrum of Casimir operators is obtained from (4.83): $C_2(\iota_1) = k^2$, $C_2'(\iota_1) = -k m_{23}$.

If the components are transformed according to (4.82), i.e. $m_{13} = j_2 m_{13}^*$, $s = m_{23} = \iota_1 j_2 j_3 m_{23}^*$, then a, b are substituted for

$$\tilde{a}(m_{12}) = is\sqrt{\frac{j_3^2(m_{13} + j_2)^2 - m_{12}^2}{m_{12}^2(4m_{12}^2 - j_2^2 j_3^2)}}, \quad \tilde{b} = \frac{s(m_{13} + j_2)}{m_{12}(m_{12} + j_2 j_3)}. \qquad (4.91)$$

The inequalities (4.88) for $j_2 = j_3 = 1$ determine: $m_{13} \geq m_{12} \geq 0$, $|m_{11}| \leq m_{12}$, $m_{11}, m_{12}, m_{13} \in \mathbb{Z}$, $s \in \mathbb{R}$. The spectrum of Casimir operators comes out of (4.82): $C_2(\iota_1) = s^2$, $C_2'(\iota_1) = -s m_{13}$.

For algebra $so(4; j_1, j_2, \iota_3) = T_3 \ni so(3; j_1, j_2)$ where $T_3 = \{X_{03}, X_{13}, X_{23}\}$ is the Abelian subalgebra, the relations (4.87) determine the operators of irreducible representations under the transformation (4.83), i.e. $k = m_{13} = j_1 j_2 \iota_3 m_{13}^*$, $m_{23} = j_2 m_{23}^*$, $p = m_{12} = j_2 \iota_3 m_{12}^*$, $q = m_{11} = \iota_3 m_{11}^*$, as follows:

$$X_{01}|m\rangle = ikq\frac{m_{23}}{p^2}|m\rangle + \frac{1}{2p}\sqrt{(p^2 - j_2^2 q^2)(k^2 - j_1^2 p^2)}$$

$$\times \left[2|m\rangle_p' + \frac{p}{p^2 - j_2^2 q^2}|m\rangle - \frac{k^2}{p(k^2 - j_1^2 p^2)}|m\rangle \right],$$

$$X_{02}|m\rangle = ik\frac{m_{23}}{p^2}\sqrt{p^2 - j_2^2 q^2}|m\rangle - \frac{1}{2p}\sqrt{k^2 - j_1^2}$$

$$\times \left[2p|m\rangle_q' + 2j_2^2 q|m\rangle_p' - \frac{j_2^2 qk^2}{p(k^2 - j_1^2 p^2)}|m\rangle \right],$$

$$X_{03}|m\rangle = i\sqrt{k^2 - j_1^2 p^2}|m\rangle, \quad |m\rangle = \begin{vmatrix} k & & m_{23} \\ & p & \\ & q & \end{vmatrix}. \qquad (4.92)$$

The rest of the operators are given by (4.75) with obvious modifications. The spectrum of Casimir operators is $C_2(\iota_3) = k^2$, $C_2'(\iota_3) = -km_{23}$. The inequalities (4.86) for $j_1 = j_2 = 1$ imply $0 \le |m_{23}| < \infty$, $k \ge p \ge 0$, $|q| \le p$, $m_{23} \in \mathbb{Z}$, $k, p, q \in \mathbb{R}$.

For the transformation of components (4.82), i.e. $m_{13} = j_2 m_{13}^*$, $s = m_{23} = j_1 j_2 \iota_3 m_{23}^*$, $p = m_{12} = j_2 \iota_3 m_{12}^*$, $q = m_{11} = \iota_3 m_{11}^*$, the representation of algebra $so(4; j_1, j_2, \iota_3)$ is described by the operators

$$X_{01}|\tilde{m}\rangle = \frac{isq}{p^2}(m_{13} + j_2)|\tilde{m}\rangle + \frac{i}{2p}\sqrt{(p^2 - j_2^2 q^2)(j_1^2 p^2 - s^2)}$$

$$\times \left[2|\tilde{m}\rangle_p' + \frac{s^2}{p(j_1^2 p^2 - s^2)}|\tilde{m}\rangle + \frac{p}{(p^2 - j_2^2 q^2)}|\tilde{m}\rangle \right],$$

$$X_{02}|\tilde{m}\rangle = \frac{is}{p^2}(m_{13} + j_2)\sqrt{p^2 - j_2^2 q^2}|\tilde{m}\rangle - \frac{i}{2p}\sqrt{j_1^2 p^2 - s^2}$$

$$\times \left[2p|\tilde{m}\rangle_q' + 2j_2^2 q|\tilde{m}\rangle_p' + \frac{j_2^2 s^2 q}{p(j_1^2 p^2 - s^2)}|\tilde{m}\rangle \right],$$

$$X_{03}|\tilde{m}\rangle = -\sqrt{j_1^2 p^2 - s^2}|\tilde{m}\rangle, \quad |\tilde{m}\rangle = \begin{vmatrix} m_{13} & & s \\ & p & \\ & q & \end{vmatrix}. \qquad (4.93)$$

The components of pattern $|\tilde{m}\rangle$ satisfy the inequalities implied by (4.82) for $j_1 = j_2 = 1$: $m_{13} \geq 0$, $p \geq |s|$, $|q| \leq p$, $m_{13} \in \mathbb{Z}$, $p, q, s \in \mathbb{R}$. The spectrum of Casimir operators is as follows: $C_2(\iota_3) = s^2$, $C_2'(\iota_3) = -s(m_{13} + j_2)$. It follows from (4.75) and (4.93) that the generators X_{13}, X_{23}, $X_{03} \in T_3$ are diagonal in the continuous basis $|\tilde{m}\rangle$.

Rejecting to fix the coordinate axes, we notice that algebra $so(4; \iota_1, 1, 1)$ is isomorphic to algebra $so(4; 1, 1, \iota_3)$, and both algebras are isomorphic to the inhomogeneous algebra $iso(3)$. Isomorphism can be established by relating the generator X_{rs}, $r < s$ to the generator $X_{3-s,3-r}$ of another algebra.

Then the operators (4.87) and (4.90) determine the irreducible representation of algebra $iso(3)$ in discrete basis corresponding to the chain of subalgebras $iso(3) \supset so(3) \supset so(2)$, and the operators (4.92) and (4.75) describe the same representation in continuous basis, corresponding to the chain $iso(3) \supset so(3; 1, \iota_3) \supset so(2; \iota_3)$. The same assertion is valid for another variant of transition from the representation of algebra $so(4)$ to the representations of algebra $so(4; j)$, which brings to (4.87), (4.91) and (4.75), (4.93). It is worth noticing that the contractions of representations give another way of constructing the irreducible representations of algebras (groups) with the structure of a semidirect sum (product).

The operators of irreducible representation of algebra $so(4; j_1, \iota_2, j_3)$ arises from (4.87) for $j_2 = \iota_2$ and can be written as follows:

$$X_{01}|m\rangle = \frac{iks}{p^2} m_{11}|m\rangle + f(k, p, s) \left[2|m\rangle_p' \right.$$
$$\left. + \frac{j_3^2 s^2}{p(p^2 - j_3^2 s^2)}|m\rangle - \frac{j_1^2 p^2}{k^2 - j_1^2 p^2}|m\rangle \right],$$

$$X_{02}|m\rangle = \frac{iks}{2p}(|m_{11} + j_3\rangle + |m_{11} - j_3\rangle)$$
$$- \frac{1}{j_3} f(k, p, s)(|m_{11} + j_3\rangle - |m_{11} - j_3\rangle),$$

$$X_{03}|m\rangle = \frac{iks}{2p}(|m_{11} + j_3\rangle - |m_{11} - j_3\rangle)$$
$$+ if(k, p, s)(|m_{11} + j_3\rangle + |m_{11} - j_3\rangle),$$

$$f(k, p, s) = \frac{1}{2p}\sqrt{(k^2 - j_1^2 p^2)(p^2 - j_3^2 s^2)}, \quad |m\rangle = \begin{vmatrix} k & & s \\ & p & \\ & m_{11} & \end{vmatrix}. \qquad (4.94)$$

The components of pattern $|m\rangle$ satisfy the inequalities: $k \geq |s|$, $-\infty < s < \infty$, $k \geq p \geq |s|$, $-\infty < m_{11} < \infty$, $k, p, s \in \mathbb{R}$, $m_{11} \in \mathbb{Z}$, if $j_1 = j_3 = 1$. The spectrum of Casimir operators is as follows: $C_2(\iota_2) = k^2 + j_1^2 j_3^2 s^2$, $C_2'(\iota_2) = -ks$.

For algebra $so(4; \iota_1, \iota_2, j_3)$ the operators of representation are given by (4.87) for $j_1 = \iota_1$, $j_2 = \iota_2$, i.e.

$$X_{01}|m\rangle = \frac{iksm_{11}}{p^2}|m\rangle + \frac{k}{2p}\sqrt{p^2 - j_3^2 s^2}\left[2|m\rangle_p' + \frac{j_3^2 s^2}{p(p^2 - j_3^2 s^2)}|m\rangle\right],$$

$$X_{02}|m\rangle = \frac{iks}{2p}(|m_{11} + j_3\rangle + |m_{11} - j_3\rangle)$$
$$- \frac{1}{j_3}\frac{k}{2p}\sqrt{p^2 - j_3^2 s^2}(|m_{11} + j_3\rangle - |m_{11} - j_3\rangle),$$

$$X_{03}|m\rangle = \frac{iks}{2p}(|m_{11} + j_3\rangle - |m_{11} - j_3\rangle)$$
$$+ \frac{ik}{2p}\sqrt{p^2 - j_3^2 s^2}(|m_{11} + j_3\rangle + |m_{11} - j_3\rangle). \tag{4.95}$$

The components of pattern $|m\rangle$ for $j_3 = 1$ satisfy the inequalities: $k \geq 0$, $-\infty < s < \infty$, $p \geq |s|$, $-\infty < m_{11} < \infty$, $k, p, s \in \mathbb{R}$, $m_{11} \in \mathbb{Z}$. The spectrum of Casimir operators is as follows: $C_2(\iota_1, \iota_2) = k^2$, $C_2'(\iota_1, \iota_2) = -ks$.

The representation of algebra $so(4; j_1, \iota_2, \iota_3)$ is given by

$$X_{01}|m\rangle = \frac{iksq}{p}|m\rangle + \frac{1}{2}\sqrt{k^2 - j_1^2 p^2}\left[2|m\rangle_p' - \frac{j_1^2 p^2}{k^2 - j_1^2 p^2}|m\rangle\right],$$

$$X_{02}|m\rangle = \frac{iks}{p}|m\rangle - \sqrt{k^2 - j_1^2 p^2}|m\rangle_q',$$

$$X_{03}|m\rangle = i\sqrt{k^2 - j_1^2 p^2}|m\rangle, \quad |m\rangle = \left|\begin{array}{ccc} k & & s \\ & p & \\ & q & \end{array}\right\rangle, \tag{4.96}$$

which arises from (4.87) for $j_2 = \iota_2$, $j_3 = \iota_3$. For $j_1 = 1$ the components of pattern $|m\rangle$ satisfy the inequalities: $k \geq 0$, $-\infty < s < \infty$, $k \geq p \geq 0$, $-\infty < q < \infty$, $k, p, s \in \mathbb{R}$. The spectrum of Casimir operators are the same as in (4.95).

Algebra $so(4; \iota_1, \iota_2, j_3) = T_5 \divideontimes so(2; j_3)$, where $so(2; j_3) = \{X_{23}\}$, is isomorphic to algebra $so(4; j_1, \iota_2, \iota_3) = T_5' \divideontimes so(2; j_1)$, $so(2; j_1) = \{X_{01}\}$, and they both are isomorphic to algebra $a = T_5 \divideontimes K$, where T_5 is a nilpotent radical, and K is a one-dimensional component subalgebra. Therefore

(4.95) determines the irreducible representation of algebra a in the basis, corresponding to the chain of subalgebras $so(4; \iota_1, \iota_2, j_3) \supset so(3; \iota_2, j_3) \supset so(2; j_3)$, where $so(2; j_3) = \{X_{23}\}$ is a compact subalgebra with discrete eigenvalues m_{11}, and (4.96) describes the same representation of algebra a in continuous basis, determined by the chain $so(4; j_1, \iota_2, \iota_3) \supset so(3; \iota_2, \iota_3) \supset so(2; \iota_3)$, where $so(2; \iota_3) = \{X_{23}\}$ and X_{23} is already a noncompact generator with continuous eigenvalues q.

For $j_1 = \iota_1$, $j_3 = \iota_3$ the formulae (4.87) give the representation of algebra $so(4; \iota_1, j_2, \iota_3)$:

$$X_{01}|m\rangle = \frac{ikqm_{23}}{p^2}|m\rangle + \frac{k}{2p}\sqrt{p^2 - j_2^2 q^2}\left[2|m\rangle'_p + \frac{j_2^2 q^2}{p(p^2 - j_2^2 q^2)}|m\rangle\right],$$

$$X_{02}|m\rangle = \frac{ikm_{23}}{p^2}\sqrt{p^2 - j_2^2 q^2}|m\rangle - \frac{k}{2p}\left[2j_2^2 q|m\rangle'_p + 2p|m\rangle'_q - j_2^2\frac{q}{p}|m\rangle\right],$$

$$X_{03}|m\rangle = ik|m\rangle, \quad |m\rangle = \begin{vmatrix} k & & m_{23} \\ & p & \\ & & q \end{vmatrix}\Big\rangle. \tag{4.97}$$

The components of pattern $|m\rangle$ for $j_2 = 1$ satisfy the following inequalities: $k \geq 0$, $-\infty < m_{23} < \infty$, $p \geq 0$, $q \leq p$, $k, p, q \in \mathbb{R}$, $m_{23} \in \mathbb{Z}$. The spectrum of Casimir operators are as follows: $C_2(\iota_1, \iota_3) = k^2$, $C_2'(\iota_1, \iota_3) = -km_{23}$.

Three-dimensional contraction $j = \iota$ turns (4.87) into the representation of maximally contracted algebra $so(4; \iota)$, described by the operators

$$X_{01}|m\rangle = \frac{iksq}{p^2}|m\rangle + k|m\rangle'_p, \quad X_{03}|m\rangle = ik|m\rangle,$$

$$X_{02}|m\rangle = \frac{iks}{p}|m\rangle - k|m\rangle'_q, \quad |m\rangle = \begin{vmatrix} k & & s \\ & p & \\ & & q \end{vmatrix}\Big\rangle, \tag{4.98}$$

with the eigenvalues of Casimir operators: $C_2(\iota) = k^2$, $C_2'(\iota) = -ks$. The components of pattern $|m\rangle$ are real, continuous and satisfy the following inequalities: $k \geq 0$, $-\infty < s < \infty$, $p \geq 0$, $-\infty < q < \infty$.

4.4.4 $so(n; j)$

In section 4.4.2 we have discussed in detail possible variants of transformations of the irreducible representation under transition from algebra $so(4)$ to algebra $so(4; j)$. For orthogonal algebras of arbitrary dimension we shall not consider all possible variants, but will concentrate on the (basic) variant where the number of parameters j, of which the components are multiplied to some row of Gel'fand–Tsetlin patterns, diminishes with increasing component number in this row. The transformation of components under transition from algebra $so(n)$ to $so(n;j)$ can be found from the rule of transformation for Casimir operators. Because orthogonal algebras of even and odd dimensions have different sets of Casimir operators, we shall discuss these cases separately.

Algebra $so(2k + 1; j)$, $j = (j_1, \ldots, j_{2k+1})$ is characterized by a set of $k + 1$ invariant operators, which are transformed according to (1.50):

$$C_{2p}(\mathbf{j}) = \prod_{s=1}^{p-1} j_s^{2s} j_{2(k+1)-s}^{2s} \prod_{l=p}^{2(k+1)-p} j_l^{2p} C_{2p}^*(\rightarrow), \quad p = 1, 2, \ldots, k,$$

$$(4.99)$$

$$C'_{k+1}(\mathbf{j}) = j_{k+1}^{k+1} \prod_{l=1}^{k} j_l^l j_{2(k+1)-l}^l C_{k+1}^{*\,'}(\rightarrow).$$

Gel'fand–Tsetlin pattern for algebra $so(2k + 2)$ is as follows:

$$|m\rangle = \begin{vmatrix} m_{1,2k+1}^* & m_{2,2k+1}^* & \cdots & m_{k,2k+1}^* & & m_{k+1,2k+1}^* \\ & m_{1,2k}^* & \cdots & \cdots & \cdots & m_{k,2k}^* \\ & m_{1,2k-1}^* & \cdots & \cdots & \cdots & m_{k,2k-1}^* \\ & & m_{1,2k-2}^* & \cdots & m_{k-1,2k-2}^* & \\ & & m_{1,2k-3}^* & \cdots & m_{k-1,2k-3}^* & \\ & & & \cdots & \cdots & \\ & & & m_{12}^* & & \\ & & & m_{11}^* & & \end{vmatrix}$$

$$(4.100)$$

The irreducible representation as well as the spectrum of Casimir operators on this representation are completely determined by the components $m_{p,2k+1}^*$ of the upper row of the pattern (4.100). The component $m_{1,2k+1}^*$ enters the spectrum of Casimir operator C_2^* quadratically, and due to this fact in the basic variant it is transformed according to the rule $m_{1,2k+1} = m_{1,2k+1}^* \prod_{l=1}^{2k+1} j_l$. The transformation of the component $m_{p,2k+1}^*$ coincides with the transformation of algebraic quantity $\sqrt{C_{2p}^*/C_{2(p-1)}^*}$:

$$m_{p,2k+1} = m_{p,2k+1}^* \prod_{l=p}^{2(k+1)-p} j_l = m_{p,2k+1}^* J_{p,2k+1}. \quad (4.101)$$

The component $m^*_{k+1,2k+1}$ is transformed in the same way as the ratio $C^{*\prime}_{k+1}/\sqrt{C^{*}_{2k}}$, i.e. $m_{k+1,2k+1} = j_{k+1}m^*_{k+1,2k+1}$. This relation, as well as the transformation relation of the component $m^*_{1,2k+1}$, is a natural result of (4.101) for $p = k+1$ and $p = 1$ respectively. Thus, all components of the highest weight (the upper row) are transformed according to (4.101). The inequalities, which govern them in the classical case, turns into

$$\frac{m_{p,2k+1}}{J_{p,2k+1}} \geq \frac{m_{p+1,2k+1}}{J_{p+1,2k+1}}, \quad \frac{m_{k,2k+1}}{J_{k,2k+1}} \geq \frac{|m_{k+1,2k+1}|}{J_{p+1,2k+1}}, \tag{4.102}$$

where $p = 1, 2, \ldots, k - 1$.

Similarly, transformation of the components of the row with number $2k$ of the pattern (4.100) is determined by the rules of transformations for Casimir operators of subalgebra $so(2k + 1; j_2, \ldots, j_{2k+1})$ and is given by (4.101), in which the product of parameters j_l starts with $p + 1$. In general the rule of transformation for the components of the pattern (4.100) can be easily found and turns out to be as follows:

$$m_{p,2s+1} = J_{p,2s+1}m^*_{p,2s+1}, \quad J_{p,2s+1} = \prod_{l=p+2(k-s)}^{2(k+1)-p} j_l,$$

$$s = 0, 1, \ldots, k, \quad p = 1, 2, \ldots, s+1,$$

$$m_{p,2s} = m^*_{p,2s}J_{p,2s}, \quad J_{p,2s} = \prod_{l=p+2(k-s)+1}^{2(k+1)-p} j_l, \tag{4.103}$$

$$s = 1, 2, \ldots, k, \quad p = 1, 2, \ldots, s.$$

The transformed components are governed by the inequalities

$$\frac{m_{p,2s+1}}{J_{p,2s+1}} \geq \frac{m_{p,2s}}{J_{p,2s}} \geq \frac{m_{p+1,2s+1}}{J_{p+1,2s+1}}, \quad p = 1, 2, \ldots, s-1,$$

$$\frac{m_{s,2s+1}}{J_{s,2s+1}} \geq \frac{m_{s,2s}}{J_{s,2s}} \geq \frac{|m_{s+1,2s+1}|}{J_{s+1,2s+1}},$$

$$\frac{m_{p,2s}}{J_{p,2s}} \geq \frac{m_{p,2s-1}}{J_{p,2s-1}} \geq \frac{m_{p+1,2s}}{J_{p+1,2s}}, \quad p = 1, 2, \ldots, s-1, \tag{4.104}$$

$$\frac{m_{s,2s}}{J_{s,2s}} \geq \frac{m_{s,2s-1}}{J_{s,2s-1}} \geq -\frac{m_{s,2s}}{J_{s,2s}},$$

which are interpreted for nilpotent and imaginary values of the parameters j according to the rules described in section 4.2.1. The action of the whole algebra $so(2k+2; j)$ can be reproduced by giving the action of the generators

$X_{2(k-s)+1,2(k-s+1)}$, $s = 1, 2, \ldots, k$, $X_{2(k-s),2(k-s)+1}$, $s = 0, 1, \ldots, k-1$.
Transforming the expressions for these generators, which can be found in
[Barut and Raczka (1977)], we obtain

$$X_{2(k-s)+1,2(k-s+1)}|m\rangle = \sum_{p=1}^{s} \frac{1}{J_{p,2s-1}} \{A(m_{p,2s-1})$$

$$|m_{p,2s-1} + J_{p,2s-1}\rangle - A(m_{p,2s-1} - J_{p,2s-1})|m_{p,2s-1} - J_{p,2s-1}\rangle\},$$

$$X_{2(k-s),2(k-s)+1}|m\rangle = iC_{2s}|m\rangle + \sum_{p=1}^{s} \frac{1}{J_{p,2s}} \{B(m_{2,2s})$$

$$|m_{p,2s} + J_{p,2s}\rangle - B(m_{p,2s} - J_{p,2s})|m_{p,2s} + J_{p,2s}\rangle\},$$

$$C_{2s} = \prod_{p=1}^{s} l_{p,2s-1} \prod_{p=1}^{s+1} l_{p,2s+1} \prod_{p=1}^{s} \frac{1}{l_{p,2s}(l_{p,2s} - J_{2,2s})},$$

$$B(m_{p,2s}) = \left\{ \prod_{r=1}^{p-1}(l_{r,2s-1}^2 - l_{p,2s}^2 a_{r,p,s}^2) \prod_{r=p}^{s}(l_{r,2s-1}^2 a_{r,p,s}^{-2} - l_{p,2s}^2) \right.$$

$$\times \prod_{r=1}^{p}(l_{r,2s+1}^2 - l_{p,2s}^2 b_{r,p,s}^2) \prod_{r=p+1}^{s+1}(l_{r,2s+1}^2 b_{r,p,s}^{-2} - l_{p,2s}^2) \bigg\}^{1/2}$$

$$\times \left\{ l_{p,2s}^2(4l_{p,2s}^2 - J_{p,2s}^2) \prod_{r=1}^{p-1}(l_{r,2s}^2 - l_{p,2s}^2 c_{r,p,s}^2)[(l_{r,2s} - J_{r,2s})^2 \right.$$

$$\left. - l_{p,2s} c_{r,p,s}^2] \prod_{r=p+1}^{s}(l_{r,2s}^2 c_{r,p,s}^{-2} - l_{p,2s}^2)[(l_{r,2s} - J_{r,2s})^2 c_{r,p,s}^{-2} - l_{p,2s}^2] \right\}^{-1/2},$$

$$A(m_{p,2s-1}) = \frac{1}{2} \left\{ \prod_{r=1}^{p-1}(l_{r,2s-2} - l_{p,2s-1}a_{r,p,s-1/2} - J_{r,2s-2}) \right.$$

$$\times (l_{2,2s-2} + l_{p,2s-1}a_{r,p,s-1/2} \prod_{r=p}^{s-1}(l_{r,2s-2}a_{r,p,s-1/2}^{-1} - l_{p,2s-1})$$

$$- J_{p,2s-1})(l_{r,2s-2}a_{r,p,s-1/2}^{-1} + l_{p,2s-1}) \prod_{r=1}^{p}(l_{r,2s}$$

$$- l_{p,2s-1}b_{r,p,s-1/2} - J_{r,2s})(l_{r,2s} + l_{p,2s-1}b_{r,p,s-1/2})$$

$$\times \prod_{r=p+1}^{s}(l_{r,2s}b_{r,p,s-1/2}^{-1} - l_{p,2s-1} - J_{p,2s-1})(l_{r,2s}b_{r,p,s-1/2}^{-1} + l_{p,2s-1}) \right\}^{1/2}$$

$$\times \left\{ \prod_{r=p+1}^{s} (l_{r,2s-1}^2 c_{r,p,s-1/2}^{-2} - l_{p,2s-1}^2)[l_{r,2s-1}^2 \right.$$

$$- (l_{p,2s-1} + J_{p,2s-1})c_{r,p,s-1/2}^2] \prod_{r=p+1}^{s} (l_{r,2s-1}^2 c_{r,p,s-1/2}^{-2} - l_{p,2s-1}^2)$$

$$\left. [l_{r,2s-1}^2 c_{r,p,s-1/2}^{-2} - (l_{p,2s-1} + J_{p,2s-1})^2] \right\}^{-1/2} \quad ,$$

$$a_{r,p,s} = \frac{J_{r,2s-1}}{J_{p,2s}} \quad b_{r,p,s} = \frac{J_{r,2s+1}}{J_{p,2s}},$$

$$c_{r,p,s} = \frac{J_{r,2s}}{J_{p,2s}}, \quad l_{p,2s} = m_{p,2s} + (s-p+1)J_{p,2s}. \tag{4.105}$$

For algebra $so(2k+1)$ Gel'fand–Tsetlin pattern $|m^*\rangle$ coincides with (4.100) with the exception of row number $2k+1$. The upper row, determining the components of the highest weight, is now the row with number $2k$; its components satisfy the inequalities $m_{1,2k}^* \geq m_{2,2k}^* \geq \ldots \geq m_{k,2k}^* \geq 0$. Under transition from classical algebra $so(2k+1)$ to algebras $so(2k+1;j)$, $j = (j_1, \ldots, j_{2k})$ the components of pattern $|m\rangle$ are transformed as follows:

$$m_{p,2s} = m_{p,2s}^* J_{p,2s}, \quad J_{p,2s} = \prod_{l=p+2(k-s)}^{2k+1-p} j_l,$$

$$m_{p,2s-1} = m_{p,2s-1}^* J_{p,2s-1}, \quad J_{p,2s-1} = \prod_{l=p+2(k-s)+1}^{2k+1-p} j_l, \tag{4.106}$$

$s = 1, 2, \ldots, k$, $p = 1, 2, \ldots, s$. We draw our attention to the fact that the lower limits in the product, determining $J_{p,2s}, J_{p,2s-1}$, have changed in comparison with (4.103). This is due to the reduction in number of parameters j by one in the case of algebra $so(2k+1;j)$ in comparison with algebra $so(2k+2;j)$. The components of the upper row in pattern $|m\rangle$ satisfy the inequalities

$$\frac{m_{1,2k}}{J_{1,2k}} \geq \frac{m_{2,2k}}{J_{2,2k}} \geq \ldots \geq \frac{m_{k-1,2k}}{J_{k-1,2k}} \geq \frac{m_{k,2k}}{J_{k,2k}} \geq 0, \tag{4.107}$$

and the other components are governed by the inequalities (4.104), of which the parameters $J_{p,2s}, J_{p,2s-1}$ are defined according to (4.106). The operators of irreducible representation of algebra $so(2k + 1; j)$ are given by (4.105) with the parameters from (4.106).

The operators (4.105) satisfy the commutation relations of algebra $so(n; j)$ because they are obtained from those of $so(n)$ by the transformations (1.30). This can be checked by straightforward calculations as well. The irreducibility of representation results from the action of raising and lowering operators on vectors of the highest and lowest weights and nonzero outcome of this action for nilpotent and imaginary values of parameters j. Though the initial representation of algebra $so(n)$ is Hermitian, the representation (4.105) in general is not as such. Requiring the fulfillment of conditions $X_{rs}^{\dagger} = -X_{rs}$, we find those values of transformed components of Gel'fand–Tsetlin pattern, for which the representation (4.105) will be Hermitian.

It is worth noticing, at last, that for the imaginary values of parameters j the relations (4.105) give representations of pseudoorthogonal algebras of different signature including discrete representations of Nikolov (1968). Analytic continuations are not considered here. We attract attention to the contractions of representations.

Chapter 5

High-temperature limit of the Standard Model

5.1 Introduction

Today the Standard Model is the only modern theory of elementary particles. It gives a good description of the experimental data and has been recently confirmed by the discovery of Higgs boson at the LHC. If we are interested in particle properties of the early Universe we need to resort to a high-temperature (high-energy) limit of the Standard Model. Standard Model is a gauge theory with its gauge group being a direct product of the simple groups $SU(3) \times SU(2) \times U(1)$. Strong interactions of quarks are described by Quantum Chromodynamics (QCD), where its gauge group is $SU(3)$ and the characteristic temperature is $0.2\,\mathrm{GeV}$. The Electroweak Model is based on the gauge group $SU(2) \times U(1)$, where $SU(2)$ is charged with the weak interactions at the characteristic temperature $100\,\mathrm{GeV}$ while $U(1)$ is connected with the long-range electromagnetic interactions. Because a photon is massless, its characteristic temperature extends to an "infinite" Planck energy. From this observation we conclude that the gauge group of the theory of elementary particles becomes simpler when the Universe's temperature increases. We suppose that this simplification is described by group contraction.

The non-relativistic limit changes not only the physics but also the corresponding theoretical tools. In particular, the Lorentz invariance group of the special relativity space-time is transformed to the Galilei group when a dimensional parameter — the velocity of light c — tends to infinity and a dimensionless parameter tends to zero $\frac{v}{c} \to 0$. E. Wigner and E. Inönü were the first who formalized this operation and called it group contraction [Inönü and Wigner (1953)]. The conception of contraction (or limit transition) has been extended to algebraic structures like quantum groups, supergroups and others, including fundamental representations of the unitary groups [Gromov (2012)]. For a symmetric physical system, the contraction of its symmetry group means a transition to some limit state. In the case of a complicated physical system, the investigation of its limit states under the limit values of some of its parameters enables us to better understand the system behavior. We discuss the modified Standard Model with the contracted gauge group mostly at the level of classical fields.

In a broad sense of the word, deformation is an operation inverse to contraction. The nontrivial deformation of some algebraic structure generally means its non-evident generalization. Quantum groups [Reshetikhin, Takhtajan and Faddeev (1989)], which are simultaneously non-commutative and non-cocommutative Hopf algebras, present a good example of similar generalization since previously Hopf algebras with only one of these properties were known. But when the contraction of some mathematical or physical structure is performed, one can reconstruct the initial structure by deformation in the narrow sense, moving back along the contraction way.

We use this method in order to re-establish the evolution of the elementary particles and their interactions in the early Universe. We base on the modern knowledge of the particle world which is concentrated in the Standard Model. For this, we investigate the high-temperature limit of the Standard Model generated by the contraction of the gauge groups $SU(2)$ and $SU(3)$ [Gromov (2015, 2016)]. Very high temperatures can exist in the early Universe after inflation and reheating on the first stages of the Hot Big Bang [Gorbunov and Rubakov (2011)]. It appears that the Standard Model Lagrangian falls into a number of terms which are distinguished by the powers of the contraction parameter $\epsilon \to 0$. As far as the temperature in the hot Universe is connected with its age, then moving forward in time, i.e. back to the high-temperature contraction, we conclude that after the creation of the Universe, the elementary particles and their interactions undergo a number of stages in their evolution from the infinite temperature state up to the Standard Model state. These stages of quark-gluon plasma formation and color symmetries restoration are distinguished by the powers of the contraction parameter and consequently by the time of its creation. From the contraction of the Standard Model we can classify the stages in time as earlier-later, but we cannot determine their absolute date. To estimate the absolute date, we use additional assumptions.

5.2 Electroweak Model

We shall follow the books [Rubakov (2002); Peskin and Schroeder (1995)] in the description of the standard Electroweak Model. This model is a gauge theory with the gauge group $SU(2) \times U(1)$ acting in boson, lepton and quark sectors. Correspondingly, its Lagrangian is the sum of boson, lepton and quark Lagrangians

$$L = L_B + L_L + L_Q. \tag{5.1}$$

Lagrangian L is taken to be invariant with respect to the action of the gauge group in the space of fundamental representation $\mathbf{C}_2$:

$$SU(2): \ \vec{z}' = G\vec{z},$$

$$\begin{pmatrix} z_1' \\ z_2' \end{pmatrix} = \begin{pmatrix} \alpha & \beta \\ -\bar{\beta} & \bar{\alpha} \end{pmatrix} \begin{pmatrix} z_1 \\ z_2 \end{pmatrix}, \quad |\alpha|^2 + |\beta|^2 = 1, \tag{5.2}$$

$$U(1): \ \vec{z}' = e^{i\omega/2}\vec{z} = e^{i\omega Y}\vec{z}, \quad \omega \in \mathbf{R}.$$

Generator Y of the group $U(1)$ is proportional to unit matrix $Y = \frac{1}{2}\mathbf{1}$. Generators of $SU(2)$,

$$T_1 = \frac{1}{2} \begin{pmatrix} 0 & 1 \\ 1 & 0 \end{pmatrix} = \frac{1}{2}\tau_1, \quad T_2 = \frac{1}{2} \begin{pmatrix} 0 & -i \\ i & 0 \end{pmatrix} = \frac{1}{2}\tau_2,$$

$$T_3 = \frac{1}{2} \begin{pmatrix} 1 & 0 \\ 0 & -1 \end{pmatrix} = \frac{1}{2}\tau_3, \tag{5.3}$$

where τ_k are the Pauli matrices, subject to commutation relations

$$[T_1, T_2] = iT_3, \quad [T_3, T_1] = iT_2, \quad [T_2, T_3] = iT_1 \tag{5.4}$$

and form Lie algebra $su(2)$.

Boson sector $L_B = L_A + L_\phi$ involves two parts: the gauge field Lagrangian

$$L_A = \frac{1}{8g^2} \text{Tr}(F_{\mu\nu})^2 - \frac{1}{4}(B_{\mu\nu})^2$$

$$= -\frac{1}{4}[(F^1_{\mu\nu})^2 + (F^2_{\mu\nu})^2 + (F^3_{\mu\nu})^2] - \frac{1}{4}(B_{\mu\nu})^2 \qquad (5.5)$$

and the matter field Lagrangian

$$L_\phi = \frac{1}{2}(D_\mu\phi)^\dagger D_\mu\phi - \frac{\lambda}{4}\left(\phi^\dagger\phi - v^2\right)^2, \qquad (5.6)$$

where $\phi = \begin{pmatrix} \phi_1 \\ \phi_2 \end{pmatrix} \in \mathbf{C}_2$ are the matter fields. The covariant derivatives are given by

$$D_\mu\phi = \partial_\mu\phi - ig\left(\sum_{k=1}^{3} T_k A^k_\mu\right)\phi - ig'Y B_\mu\phi, \qquad (5.7)$$

where $T_k = \frac{1}{2}\tau_k, k = 1,2,3$ are the generators of $SU(2)$, $Y = \frac{1}{2}\mathbf{1}$ is the generator of $U(1)$, g and g' are constants. The gauge fields

$$A_\mu(x) = -ig\sum_{k=1}^{3} T_k A^k_\mu(x), \quad B_\mu(x) = -ig'B_\mu(x) \qquad (5.8)$$

take their values in Lie algebras $su(2)$, $u(1)$ respectively, and the stress tensors are as follows:

$$F_{\mu\nu}(x) = \mathcal{F}_{\mu\nu}(x) + [A_\mu(x), A_\nu(x)], \quad B_{\mu\nu} = \partial_\mu B_\nu - \partial_\nu B_\mu. \qquad (5.9)$$

To generate mass for the vector bosons the special mechanism of spontaneous symmetry breaking is used. One of the L_B ground states

$$\phi^{vac} = \begin{pmatrix} 0 \\ v \end{pmatrix}, \quad A^k_\mu = B_\mu = 0 \qquad (5.10)$$

is taken as a vacuum state of the model, and small field excitations $v + \chi(x)$ with respect to this vacuum are regarded.

After spontaneous symmetry breaking, the boson Lagrangian (5.5), (5.6) can be represented in the form

$$L_B = L_B^{(2)} + L_B^{int}$$

$$= \frac{1}{2}(\partial_\mu\chi)^2 - \frac{1}{2}m^2_\chi\chi^2 + \frac{1}{2}m^2_Z Z_\mu Z_\mu - \frac{1}{4}Z_{\mu\nu}Z_{\mu\nu}$$

$$- \frac{1}{4}\mathcal{F}_{\mu\nu}\mathcal{F}_{\mu\nu} + m^2_W W^+_\mu W^-_\mu - \frac{1}{2}W^+_{\mu\nu}W^-_{\mu\nu} + L_B^{int}, \qquad (5.11)$$

where as usual the second order terms describe the boson particles content of the model and higher order terms L_B^{int} are regarded as their interactions. So Lagrangian (5.11) includes charged W-bosons with identical masses $m_W = \frac{1}{2}gv$, massless photon A_μ, neutral Z-boson with mass $m_Z = \frac{v}{2}\sqrt{g^2 + g'^2}$ and scalar Higgs boson χ with $m_\chi = \sqrt{2\lambda}v$. All these particles are experimentally detected and have the masses: $m_W = 80\,\text{GeV}$, $m_Z = 91\,\text{GeV}$, $m_\chi = 125\,\text{GeV}$.

The interaction Lagrangian L_B^{int} looks as follows:

$$
\begin{aligned}
L_B^{int} ={}& \frac{gm_z}{2\cos\theta_W}(Z_\mu)^2\chi - \lambda v\chi^3 + \frac{g^2}{8\cos^2\theta_W}(Z_\mu)^2\chi^2 - \frac{\lambda}{4}\chi^4 \\
& - \frac{1}{2}\mathcal{W}_{\mu\nu}^+\mathcal{W}_{\mu\nu}^- + m_W^2 W_\mu^+ W_\mu^- \\
& - 2ig(W_\mu^+ W_\nu^- - W_\mu^- W_\nu^+)(\mathcal{F}_{\mu\nu}\sin\theta_W + \mathcal{Z}_{\mu\nu}\cos\theta_W) \\
& - \frac{i}{2}e[A_\mu(\mathcal{W}_{\mu\nu}^+ W_\nu^- - \mathcal{W}_{\mu\nu}^- W_\nu^+) - A_\nu(\mathcal{W}_{\mu\nu}^+ W_\mu^- - \mathcal{W}_{\mu\nu}^- W_\mu^+)] \\
& + gW_\mu^+ W_\mu^-\chi - \frac{i}{2}g\cos\theta_W[Z_\mu(\mathcal{W}_{\mu\nu}^+ W_\nu^- - \mathcal{W}_{\mu\nu}^- W_\nu^+) \\
& - Z_\nu(\mathcal{W}_{\mu\nu}^+ W_\mu^- - \mathcal{W}_{\mu\nu}^- W_\mu^+)] + \frac{g^2}{4}(W_\mu^+ W_\nu^- - W_\mu^- W_\nu^+)^2 \\
& + \frac{g^2}{4}W_\mu^+ W_\nu^-\chi^2 - \frac{e^2}{4}\{[(W_\mu^+)^2 + (W_\mu^-)^2](A_\nu)^2 \\
& - 2(W_\mu^+ W_\nu^+ + W_\mu^- W_\nu^-)A_\mu A_\nu + [(W_\nu^+)^2 + (W_\nu^-)^2](A_\mu)^2\} \\
& - \frac{g^2}{4}\cos\theta_W\{[(W_\mu^+)^2 + (W_\mu^-)^2](Z_\nu)^2 \\
& - 2(W_\mu^+ W_\nu^+ + W_\mu^- W_\nu^-)Z_\mu Z_\nu + [(W_\nu^+)^2 + (W_\nu^-)^2](Z_\mu)^2\} \\
& - eg\cos\theta_W\Big\{W_\mu^+ W_\mu^- A_\nu Z_\nu + W_\nu^+ W_\nu^- A_\mu Z_\mu \\
& - \frac{1}{2}(W_\mu^+ W_\nu^- + W_\nu^+ W_\mu^-)(A_\mu Z_\nu + A_\nu Z_\mu)\Big\}.
\end{aligned}
\tag{5.12}
$$

The fermion sector is represented by the lepton L_L and quark L_Q Lagrangians. The lepton Lagrangian is taken in the form

$$
L_L = L_l^\dagger i\tilde{\tau}_\mu D_\mu L_l + e_r^\dagger i\tau_\mu D_\mu e_r - h_e[e_r^\dagger(\phi^\dagger L_l) + (L_l^\dagger\phi)e_r],
\tag{5.13}
$$

where $L_l = \begin{pmatrix} \nu_l \\ e_l \end{pmatrix}$ is the $SU(2)$-doublet, e_r is the $SU(2)$-singlet (or a scalar relative to $\mathbf{C}_2$), h_e is constant, $\tau_0 = \tilde{\tau}_0 = \mathbf{1}$, $\tilde{\tau}_k = -\tau_k$, τ_μ are Pauli matrices and e_r, e_l, ν_l are two-component Lorentz spinors. The last term with the factor h_e represents electron mass. The covariant derivatives are given by the formulae:

$$D_\mu L_l = \partial_\mu L_l - i\frac{g}{\sqrt{2}} \left(W_\mu^+ T_+ + W_\mu^- T_- \right) L_l$$

$$- i\frac{g}{\cos\theta_w} Z_\mu \left(T_3 - Q\sin^2\theta_w \right) L_l - ie A_\mu Q L_l, \qquad (5.14)$$

$$D_\mu e_r = \partial_\mu e_r - ig' Q A_\mu e_r \cos\theta_w + ig' Q Z_\mu e_r \sin\theta_w,$$

where $T_\pm = T_1 \pm iT_2$, $Q = Y + T_3$ is the generator of electromagnetic subgroup $U(1)_{em}$, $Y = \frac{1}{2}\mathbf{1}$ is the hypercharge, $e = gg'(g^2 + g'^2)^{-\frac{1}{2}}$ is the electron charge and $\sin\theta_w = eg^{-1}$. The new gauge fields

$$Z_\mu = \frac{1}{\sqrt{g^2 + g'^2}} (g A_\mu^3 - g' B_\mu), \quad A_\mu = \frac{1}{\sqrt{g^2 + g'^2}} (g' A_\mu^3 + g B_\mu),$$

$$W_\mu^\pm = \frac{1}{\sqrt{2}} \left(A_\mu^1 \mp iA_\mu^2 \right) \qquad (5.15)$$

are introduced instead of (5.8).

The lepton Lagrangian (5.13) in terms of electron and neutrino fields takes the form

$$L_L = e_l^\dagger i\tilde{\tau}_\mu \partial_\mu e_l + e_r^\dagger i\tau_\mu \partial_\mu e_r - m_e(e_r^\dagger e_l + e_l^\dagger e_r)$$

$$+ \frac{g\cos 2\theta_w}{2\cos\theta_w} e_l^\dagger \tilde{\tau}_\mu Z_\mu e_l - ee_l^\dagger \tilde{\tau}_\mu A_\mu e_l g' \cos\theta_w e_r^\dagger \tau_\mu A_\mu e_r$$

$$+ g'\sin\theta_w e_r^\dagger \tau_\mu Z_\mu e_r + \nu_l^\dagger i\tilde{\tau}_\mu \partial_\mu \nu_l + \frac{g}{2\cos\theta_w} \nu_l^\dagger \tilde{\tau}_\mu Z_\mu \nu_l$$

$$+ \frac{g}{\sqrt{2}} [\nu_l^\dagger \tilde{\tau}_\mu W_\mu^+ e_l + e_l^\dagger \tilde{\tau}_\mu W_\mu^- \nu_l]. \qquad (5.16)$$

The quark Lagrangian is given by

$$L_Q = Q_l^\dagger i\tilde{\tau}_\mu D_\mu Q_l + u_r^\dagger i\tau_\mu D_\mu u_r + d_r^\dagger i\tau_\mu D_\mu d_r$$

$$- h_d[d_r^\dagger(\phi^\dagger Q_l) + (Q_l^\dagger \phi)d_r] - h_u[u_r^\dagger(\tilde{\phi}^\dagger Q_l) + (Q_l^\dagger \tilde{\phi})u_r], \qquad (5.17)$$

where the left quark fields form the $SU(2)$-doublet $Q_l = \begin{pmatrix} u_l \\ d_l \end{pmatrix}$, the right quark fields u_r, d_r are the $SU(2)$-singlets, $\tilde{\phi}_i = \epsilon_{ik}\bar{\phi}_k, \epsilon_{00} = 1, \epsilon_{ii} = -1$ is the conjugate representation of $SU(2)$ group and h_u, h_d are constants. All fields u_l, d_l, u_r, d_r are two-component Lorentz spinors. The last four terms with the factors h_d and h_u specify the d- and u-quark masses. The covariant derivatives of the quark fields are given by

$$D_\mu Q_l = \left(\partial_\mu - ig \sum_{k=1}^{3} \frac{\tau_k}{2} A_\mu^k - ig' \frac{1}{6} B_\mu \right) Q_l,$$

$$(5.18)$$

$$D_\mu u_r = \left(\partial_\mu - ig' \frac{2}{3} B_\mu \right) u_r, \quad D_\mu d_r = \left(\partial_\mu + ig' \frac{1}{3} B_\mu \right) d_r.$$

The quark Lagrangian (5.17) in terms of u- and d-quarks fields can be written as

$$L_Q = d_l^\dagger i \tilde{\tau}_\mu \partial_\mu d_l + d_r^\dagger i \tau_\mu \partial_\mu d_r - m_d(d_r^\dagger d_l + d_l^\dagger d_r) - \frac{e}{3} d_l^\dagger \tilde{\tau}_\mu A_\mu d_l$$

$$- \frac{g}{\cos\theta_w} \left(\frac{1}{2} - \frac{2}{3} \sin^2\theta_w \right) d_l^\dagger \tilde{\tau}_\mu Z_\mu d_l - \frac{1}{3} g' \cos\theta_w d_r^\dagger \tau_\mu A_\mu d_r$$

$$+ \frac{1}{3} g' \sin\theta_w d_r^\dagger \tau_\mu Z_\mu d_r + u_l^\dagger i \tilde{\tau}_\mu \partial_\mu u_l + u_r^\dagger i \tau_\mu \partial_\mu u_r - m_u(u_r^\dagger u_l + u_l^\dagger u_r)$$

$$+ \frac{g}{\cos\theta_w} \left(\frac{1}{2} - \frac{2}{3} \sin^2\theta_w \right) u_l^\dagger \tilde{\tau}_\mu Z_\mu u_l + \frac{2e}{3} u_l^\dagger \tilde{\tau}_\mu A_\mu u_l$$

$$+ \frac{g}{\sqrt{2}} \left[u_l^\dagger \tilde{\tau}_\mu W_\mu^+ d_l + d_l^\dagger \tilde{\tau}_\mu W_\mu^- u_l \right] + \frac{2}{3} g' \cos\theta_w u_r^\dagger \tau_\mu A_\mu u_r$$

$$- \frac{2}{3} g' \sin\theta_w u_r^\dagger \tau_\mu Z_\mu u_r, \qquad (5.19)$$

where $m_e = h_e v/\sqrt{2}$ and $m_u = h_u v/\sqrt{2}$, $m_d = h_d v/\sqrt{2}$ represent electron and quark masses.

From the viewpoint of the electroweak interactions all known leptons and quarks are divided into three generations. The next two lepton generations are introduced in a similar way to (5.13). They are left $SU(2)$-doublets

$$\begin{pmatrix} \nu_\mu \\ \mu \end{pmatrix}_l, \quad \begin{pmatrix} \nu_\tau \\ \tau \end{pmatrix}_l, \quad Y = -\frac{1}{2} \qquad (5.20)$$

and right $SU(2)$-singlets: $\mu_r, \tau_r, Y = -1$. In addition to u- and d-quarks of the first generation, there is (c, s) and (t, b) quarks of the next generations, whose left fields

$$\begin{pmatrix} c_l \\ s_l \end{pmatrix}, \quad \begin{pmatrix} t_l \\ b_l \end{pmatrix}, \quad Y = \frac{1}{6} \qquad (5.21)$$

are described by the $SU(2)$-doublets and the right fields are $SU(2)$-singlets: c_r, t_r, $Y = \frac{2}{3}$; s_r, b_r, $Y = -\frac{1}{3}$. Their Lagrangians are introduced in a similar way to (5.17). Full lepton and quark Lagrangians are obtained by the summation over all generations. In what follows we will discuss only the first generations of leptons and quarks.

5.3 High-temperature Lagrangian of EWM

We consider a model where the contracted gauge group $SU(2;\epsilon) \times U(1)$ acts in the boson, lepton and quark sectors. We introduce the contraction parameter ϵ and *consistent rescaling* of the fundamental representation of the group $SU(2)$ and the space $\mathbf{C}_2$ as follows

$$z'(\epsilon) = \begin{pmatrix} z'_1 \\ \epsilon z'_2 \end{pmatrix} = \begin{pmatrix} \alpha & \epsilon\beta \\ -\epsilon\bar{\beta} & \bar{\alpha} \end{pmatrix} \begin{pmatrix} z_1 \\ \epsilon z_2 \end{pmatrix} = u(\epsilon)z(\epsilon),$$

$$\det u(\epsilon) = |\alpha|^2 + \epsilon^2|\beta|^2 = 1, \quad u(\epsilon)u^\dagger(\epsilon) = 1,$$
(5.22)

where the real contraction parameter tends to zero $\epsilon \to 0$. The Hermite form

$$z^\dagger(\epsilon)z(\epsilon) = |z_1|^2 + \epsilon^2|z_2|^2$$
(5.23)

remain invariant under the contraction limit.

The contracted group $SU(2;\epsilon = 0)$ is isomorphic to the Euclid group $E(2)$. The space $\mathbf{C}_2(\epsilon = 0)$ is split in the limit $\epsilon \to 0$ on the one-dimensional base $\{z_1\}$ and the one-dimensional fiber $\{z_2\}$. (A simple and the best known example of a fiber space is the non-relativistic space-time with one-dimensional base, which is interpreted as time, and three-dimensional fiber, which is interpreted as a proper space.) The actions of the unitary group $U(1)$ and the electromagnetic subgroup $U(1)_{em}$ in the space $\mathbf{C}_2(\epsilon = 0)$ are given by the same matrices as in $\mathbf{C}_2$.

The space $\mathbf{C}_2(\epsilon)$ of the fundamental representation of $SU(2;\epsilon)$ group can be obtained from $\mathbf{C}_2$ by substitution of z_2 with ϵz_2. This substitution induces the other ones for Lie algebra generators $T_1 \to \epsilon T_1, T_2 \to \epsilon T_2, T_3 \to T_3$, with the new comutation relations

$$[T_1, T_2] = i\epsilon^2 T_3, \quad [T_3, T_1] = iT_2, \quad [T_2, T_3] = iT_1$$
(5.24)

for Lie algebra $su(2; \epsilon)$. The contracted algebra $su(2; \epsilon = 0)$ has the structure of a semi-direct sum of Abelian subalgebra $t_2 = \{T_1, T_2\}$ and one-dimensional subalgebra $u(1) = \{T_3\} : su(2; \epsilon) = t_2 \uplus u(1)$.

As far as the gauge fields take their values in Lie algebra, we can substitute the gauge fields instead of transforming the generators, namely:

$$A_\mu^1 \to \epsilon A_\mu^1, \quad A_\mu^2 \to \epsilon A_\mu^2, \quad A_\mu^3 \to A_\mu^3, \quad B_\mu \to B_\mu. \qquad (5.25)$$

Indeed, due to commutativity and associativity of multiplication by ϵ

$$su(2; \epsilon) \ni \{A_\mu^1(\epsilon T_1) + A_\mu^2(\epsilon T_2) + A_\mu^3 T_3\}$$
$$= \{(\epsilon A_\mu^1)T_1 + (\epsilon A_\mu^2)T_2 + A_\mu^3 T_3\}. \qquad (5.26)$$

For the standard gauge fields (5.15) these substitutions are as follows:

$$W_\mu^\pm \to \epsilon W_\mu^\pm, \quad Z_\mu \to Z_\mu, \quad A_\mu \to A_\mu. \qquad (5.27)$$

The left lepton $L_l = \begin{pmatrix} \nu_l \\ e_l \end{pmatrix}$, and quark $Q_l = \begin{pmatrix} u_l \\ d_l \end{pmatrix}$ fermionic fields are $SU(2)$-doublets, so their components are transformed in a similar way as the components of the vector z, namely:

$$e_l \to \epsilon e_l, \quad d_l \to \epsilon d_l, \quad \nu_l \to \nu_l, \quad u_l \to u_l. \qquad (5.28)$$

The right lepton and quark fields are $SU(2)$-singlets and therefore remain unchanged.

The next reason for inequality of the first and second doublet components is the special mechanism of spontaneous symmetry breaking, which is used to generate mass of vector bosons and other elementary particles of the model. In this mechanism, one of the Lagrangian L_B ground states $\phi^{vac} = \begin{pmatrix} 0 \\ v \end{pmatrix}$ is taken as vacuum of the model and then small field excitations $v + \chi(x)$ with respect to the second component of the vacuum vector are regarded. So Higgs boson field χ and constant v are multiplied by ϵ. As far as masses of all particles are proportionate to v we obtain the following transformation rule

$$\chi \to \epsilon \chi, \quad v \to \epsilon v, \quad m_p \to \epsilon m_p, \quad p = \chi, W, Z, e, u, d. \qquad (5.29)$$

After transformations (5.27)–(5.29) and spontaneous symmetry breaking, the boson Lagrangian (5.5), (5.6) can be represented in the form

$$L_B(\epsilon) = -\frac{1}{4}\mathcal{Z}_{\mu\nu}^2 - \frac{1}{4}\mathcal{F}_{\mu\nu}^2 + \epsilon^2 L_{B,2} + \epsilon^3 g W_\mu^+ W_\mu^- \chi + \epsilon^4 L_{B,4}, \qquad (5.30)$$

where

$$L_{B,4} = m_W^2 W_\mu^+ W_\mu^- - \frac{1}{2}m_\chi^2\chi^2 - \lambda v\chi^3 - \frac{\lambda}{4}\chi^4$$

$$+ \frac{g^2}{4}(W_\mu^+ W_\nu^- - W_\mu^- W_\nu^+)^2 + \frac{g^2}{4}W_\mu^+ W_\nu^-\chi^2, \qquad (5.31)$$

$$L_{B,2} = \frac{1}{2}(\partial_\mu\chi)^2 + \frac{1}{2}m_Z^2(Z_\mu)^2 - \frac{1}{2}\mathcal{W}_{\mu\nu}^+\mathcal{W}_{\mu\nu}^-$$

$$+ \frac{gm_z}{2\cos\theta_W}(Z_\mu)^2\chi + \frac{g^2}{8\cos^2\theta_W}(Z_\mu)^2\chi^2$$

$$- 2ig(W_\mu^+ W_\nu^- - W_\mu^- W_\nu^+)(\mathcal{F}_{\mu\nu}\sin\theta_W + \mathcal{Z}_{\mu\nu}\cos\theta_W)$$

$$- \frac{i}{2}e\left[A_\mu(\mathcal{W}_{\mu\nu}^+ W_\nu^- - \mathcal{W}_{\mu\nu}^- W_\nu^+) + \frac{i}{2}eA_\nu(\mathcal{W}_{\mu\nu}^+ W_\mu^- - \mathcal{W}_{\mu\nu}^- W_\mu^+)\right]$$

$$- \frac{i}{2}g\cos\theta_W[Z_\mu(\mathcal{W}_{\mu\nu}^+ W_\nu^- - \mathcal{W}_{\mu\nu}^- W_\nu^+)$$

$$- Z_\nu(\mathcal{W}_{\mu\nu}^+ W_\mu^- - \mathcal{W}_{\mu\nu}^- W_\mu^+)] - \frac{e^2}{4}\{[(W_\mu^+)^2 + (W_\mu^-)^2](A_\nu)^2$$

$$- 2(W_\mu^+ W_\nu^+ + W_\mu^- W_\nu^-)A_\mu A_\nu + [(W_\nu^+)^2 + (W_\nu^-)^2](A_\mu)^2\}$$

$$- \frac{g^2}{4}\cos\theta_W\{[(W_\mu^+)^2 + (W_\mu^-)^2](Z_\nu)^2$$

$$- 2(W_\mu^+ W_\nu^+ + W_\mu^- W_\nu^-)Z_\mu Z_\nu + [(W_\nu^+)^2 + (W_\nu^-)^2](Z_\mu)^2\}$$

$$- eg\cos\theta_W\left[W_\mu^+ W_\mu^- A_\nu Z_\nu + W_\nu^+ W_\nu^- A_\mu Z_\mu\right.$$

$$\left. - \frac{1}{2}(W_\mu^+ W_\nu^- + W_\nu^+ W_\mu^-)(A_\mu Z_\nu + A_\nu Z_\mu)\right]. \qquad (5.32)$$

The lepton Lagrangian (5.13), (5.16) in terms of electron and neutrino fields takes the form

$$L_L(\epsilon) = L_{L,0} + \epsilon^2 L_{L,2}$$
$$= \nu_l^\dagger i\tilde{\tau}_\mu \partial_\mu \nu_l + e_r^\dagger i\tilde{\tau}_\mu \partial_\mu e_r + g' \sin\theta_w e_r^\dagger \tilde{\tau}_\mu Z_\mu e_r$$
$$- g' \cos\theta_w e_r^\dagger \tilde{\tau}_\mu A_\mu e_r + \frac{g}{2\cos\theta_w} \nu_l^\dagger \tilde{\tau}_\mu Z_\mu \nu_l$$
$$+ \epsilon^2 \Big\{ e_l^\dagger i\tilde{\tau}_\mu \partial_\mu e_l - m_e(e_r^\dagger e_l + e_l^\dagger e_r) + \frac{g\cos 2\theta_w}{2\cos\theta_w} e_l^\dagger \tilde{\tau}_\mu Z_\mu e_l$$
$$- e e_l^\dagger \tilde{\tau}_\mu A_\mu e_l + \frac{g}{\sqrt{2}} \left(\nu_l^\dagger \tilde{\tau}_\mu W_\mu^+ e_l + e_l^\dagger \tilde{\tau}_\mu W_\mu^- \nu_l \right) \Big\}. \tag{5.33}$$

The quark Lagrangian (5.17), (5.19) in terms of u- and d-quarks fields can be written as

$$L_Q(\epsilon) = L_{Q,0} - \epsilon\, m_u(u_r^\dagger u_l + u_l^\dagger u_r) + \epsilon^2 L_{Q,2}, \tag{5.34}$$

where

$$L_{Q,0} = d_r^\dagger i\tilde{\tau}_\mu \partial_\mu d_r + u_l^\dagger i\tilde{\tau}_\mu \partial_\mu u_l + u_r^\dagger i\tilde{\tau}_\mu \partial_\mu u_r$$
$$- \frac{1}{3} g' \cos\theta_w d_r^\dagger \tilde{\tau}_\mu A_\mu d_r + \frac{1}{3} g' \sin\theta_w d_r^\dagger \tilde{\tau}_\mu Z_\mu d_r + \frac{2e}{3} u_l^\dagger \tilde{\tau}_\mu A_\mu u_l$$
$$+ \frac{g}{\cos\theta_w} \left(\frac{1}{2} - \frac{2}{3}\sin^2\theta_w \right) u_l^\dagger \tilde{\tau}_\mu Z_\mu u_l$$
$$+ \frac{2}{3} g' \cos\theta_w u_r^\dagger \tilde{\tau}_\mu A_\mu u_r - \frac{2}{3} g' \sin\theta_w u_r^\dagger \tilde{\tau}_\mu Z_\mu u_r, \tag{5.35}$$
$$L_{Q,2} = d_l^\dagger i\tilde{\tau}_\mu \partial_\mu d_l - m_d(d_r^\dagger d_l + d_l^\dagger d_r) - \frac{e}{3} d_l^\dagger \tilde{\tau}_\mu A_\mu d_l$$
$$- \frac{g}{\cos\theta_w} \left(\frac{1}{2} - \frac{2}{3}\sin^2\theta_w \right) d_l^\dagger \tilde{\tau}_\mu Z_\mu d_l$$
$$+ \frac{g}{\sqrt{2}} \left[u_l^\dagger \tilde{\tau}_\mu W_\mu^+ d_l + d_l^\dagger \tilde{\tau}_\mu W_\mu^- u_l \right]. \tag{5.36}$$

The complete Lagrangian of the modified model is given by the sum $L(\epsilon) = L_B(\epsilon) + L_L(\epsilon) + L_Q(\epsilon)$ and can be written in the form

$$L(\epsilon) = L_\infty + \epsilon L_1 + \epsilon^2 L_2 + \epsilon^3 L_3 + \epsilon^4 L_4. \tag{5.37}$$

The contraction parameter is a monotonous function $\epsilon(T)$ of the temperature (or average energy) with the property $\epsilon(T) \to 0$ for $T \to \infty$. Very high temperatures can exist in the early Universe just after its creation.

It is well known that to gain a better understanding of a physical system it is useful to investigate its properties for the limiting values of physical parameters. It follows from (5.37) that there are five stages in the evolution of the Electroweak Model after the creation of the Universe which are distinguished by the powers of the contraction parameter ϵ. This offers an opportunity for construction of intermediate limit models. One can take the Lagrangian L_∞ for the initial limit system, then add L_1 and obtain the second limit model with the Lagrangian $\mathcal{L}_1 = L_\infty + L_1$. After that, one can add L_2 and obtain the third limit model $\mathcal{L}_2 = L_\infty + L_1 + L_2$. The last limit model has the Lagrangian $\mathcal{L}_3 = L_\infty + L_1 + L_2 + L_3$. But it should be noted that among all limit models, only L_∞ is the gauge model with the gauge group isomorphic to Euclid group $E(2)$.

In the infinite temperature limit ($\epsilon = 0$) Lagrangian (5.37) is equal to

$$
L_\infty = -\frac{1}{4}\mathcal{Z}_{\mu\nu}^2 - \frac{1}{4}\mathcal{F}_{\mu\nu}^2 + \nu_l^\dagger i\tilde{\tau}_\mu \partial_\mu \nu_l + u_l^\dagger i\tilde{\tau}_\mu \partial_\mu u_l
$$
$$
+ e_r^\dagger i\tilde{\tau}_\mu \partial_\mu e_r + d_r^\dagger i\tilde{\tau}_\mu \partial_\mu d_r + u_r^\dagger i\tilde{\tau}_\mu \partial_\mu u_r + L_\infty^{int}(A_\mu, Z_\mu), \qquad (5.38)
$$

where

$$
L_\infty^{int}(A_\mu, Z_\mu) = \frac{g}{2\cos\theta_w}\nu_l^\dagger \tilde{\tau}_\mu Z_\mu \nu_l + \frac{2e}{3}u_l^\dagger \tilde{\tau}_\mu A_\mu u_l + g'\sin\theta_w e_r^\dagger \tau_\mu Z_\mu e_r
$$
$$
+ \frac{g}{\cos\theta_w}\left(\frac{1}{2} - \frac{2}{3}\sin^2\theta_w\right) u_l^\dagger \tilde{\tau}_\mu Z_\mu u_l - g'\cos\theta_w e_r^\dagger \tau_\mu A_\mu e_r
$$
$$
- \frac{1}{3}g'\cos\theta_w d_r^\dagger \tau_\mu A_\mu d_r + \frac{1}{3}g'\sin\theta_w d_r^\dagger \tau_\mu Z_\mu d_r
$$
$$
+ \frac{2}{3}g'\cos\theta_w u_r^\dagger \tau_\mu A_\mu u_r - \frac{2}{3}g'\sin\theta_w u_r^\dagger \tau_\mu Z_\mu u_r. \qquad (5.39)
$$

We can conclude that the limit model includes only *massless particles*: photons A_μ and neutral bosons Z_μ, left quarks u_l and neutrinos ν_l, right electrons e_r and quarks u_r, d_r. This phenomenon has a simple physical explanation: the temperature is so high, that the particle mass becomes a negligible quantity as compared to its kinetic energy. The electroweak interactions become long-range because they are mediated by the massless neutral Z-bosons and photons. Let us note that $W_\mu^\pm$-boson fields, which correspond to the translation subgroup of Euclid group $E(2)$, are absent in the limit Lagrangians (5.38), (5.39).

Similar high energies can exist in the early Universe after inflation and reheating on the first stages of the Hot Big Bang [Gorbunov and Rubakov (2011); Linde (1990)] in the pre-electroweak epoch, as it is shown

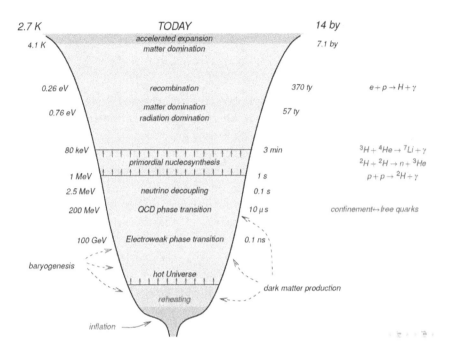

Fig. 5.1   Evolution of the Universe ($1eV = 10^4 K$).

in Fig. 5.1. However, the Universe evolution is more interesting and the limit Lagrangian L_∞ can be considered as a good approximation after the Big Bang, just as the non-relativistic mechanics is a good approximation of the relativistic one at low velocities.

From the explicit form of the interaction part $L_\infty^{int}(A_\mu, Z_\mu)$ it follows that there are no interactions between particles of different kind; for example, neutrinos interact only with each other by neutral currents. All other particles are charged and interact with particles of the same sort by massless Z_μ-bosons and photons. Particles of different kind do not interact. It looks like some stratification of the Electroweak Model with only one sort of particles in each stratum.

From the contraction of the Electroweak Model we can classify events in time as earlier-later, but we cannot determine their absolute time without additional assumptions. At the level of classical gauge fields we can already conclude that the u-quark first restores its mass in the evolution of the Universe. Indeed the mass term of the u-quark in the Lagrangian (5.37) $L_1 = -m_u(u_r^\dagger u_l + u_l^\dagger u_r)$ is proportional to the first power ϵL_1, whereas the mass terms of Z-boson, electron and d-quark are multiplied by the second power of the contraction parameter

$$\epsilon^2 \left[\frac{1}{2} m_Z^2 \left(Z_\mu \right)^2 + m_e(e_r^\dagger e_l + e_l^\dagger e_r) + m_d(d_r^\dagger d_l + d_l^\dagger d_r) \right]. \qquad (5.40)$$

At the same time, massless Higgs boson χ and charged W-boson appear. They restore their masses after all other particles of the Electroweak Model because their mass terms are multiplied by ϵ^4.

The electroweak interactions between the elementary particles are restored mainly in the epoch which corresponds to the second order of the contraction parameter. There is one term in Lagrangian (5.30) $L_3 = gW_\mu^+ W_\mu^- \chi$ proportionate to ϵ^3. The final reconstruction of the electroweak interactions takes place at the last stage ($\approx \epsilon^4$) together with the restoration of mass of all particles.

Two other generations of leptons (5.20) and quarks (5.21) develop in a similar way: for the infinite energy there are only massless right μ- and τ-muons, left μ- and τ-neutrinos, as well as massless left and right quarks $c_l, c_r, s_r, t_l, t_r, b_r$. c- and t-quarks first acquire their mass and after that μ-, τ-muons, s-, b-quarks become massive.

5.4   Lagrangian of Quantum Chromodynamics

Strong interactions of quarks are described by the Quantum Chromodynamics (QCD). Like the Electroweak Model, QCD is a gauge theory based on the local color degrees of freedom [Emel'yanov (2007)]. The QCD gauge group is $SU(3)$, acting in three dimensional complex space $\mathbf{C}_3$ of color quark states $q = \begin{pmatrix} q_1 \\ q_2 \\ q_3 \end{pmatrix} \equiv \begin{pmatrix} q_R \\ q_G \\ q_B \end{pmatrix} \in \mathbf{C}_3$, where $q(x)$ are quark fields $q = u, d, s, c, b, t$ and R (red), G (green), B (blue) are color degrees of freedom. The $SU(3)$ gauge bosons are called gluons. There are eight gluons in total, which are the force carrier of the theory between quarks. The QCD Lagrangian is taken in the form

$$\mathcal{L} = \sum_q \bar{q}^i (i\gamma^\mu)(D_\mu)_{ij} q^j - \frac{1}{4} \sum_{\alpha=1}^{8} F_{\mu\nu}^\alpha F^{\mu\nu\,\alpha}, \qquad (5.41)$$

where $D_\mu q$ are covariant derivatives of quark fields

$$D_\mu q = \left(\partial_\mu - i g_s \left(\frac{\lambda^\alpha}{2} \right) A_\mu^\alpha \right) q, \qquad (5.42)$$

g_s is the strong coupling constant, $t^a = \lambda^a/2$ are the generators of $SU(3)$, λ^a are Gell-Mann matrices in the form

$$\lambda^1 = \begin{pmatrix} \cdot & 1 & \cdot \\ 1 & \cdot & \cdot \\ \cdot & \cdot & \cdot \end{pmatrix}, \quad \lambda^2 = \begin{pmatrix} \cdot & -i & \cdot \\ i & \cdot & \cdot \\ \cdot & \cdot & \cdot \end{pmatrix}, \quad \lambda^3 = \begin{pmatrix} 1 & \cdot & \cdot \\ \cdot & -1 & \cdot \\ \cdot & \cdot & \cdot \end{pmatrix},$$

$$\lambda^4 = \begin{pmatrix} \cdot & \cdot & 1 \\ \cdot & \cdot & \cdot \\ 1 & \cdot & \cdot \end{pmatrix}, \quad \lambda^5 = \begin{pmatrix} \cdot & \cdot & -i \\ \cdot & \cdot & \cdot \\ i & \cdot & \cdot \end{pmatrix}, \quad \lambda^6 = \begin{pmatrix} \cdot & \cdot & \cdot \\ \cdot & \cdot & 1 \\ \cdot & 1 & \cdot \end{pmatrix}, \quad (5.43)$$

$$\lambda^7 = \begin{pmatrix} \cdot & \cdot & \cdot \\ \cdot & \cdot & -i \\ \cdot & i & \cdot \end{pmatrix}, \quad \lambda^8 = \frac{1}{\sqrt{3}}\begin{pmatrix} 1 & \cdot & \cdot \\ \cdot & 1 & \cdot \\ \cdot & \cdot & -2 \end{pmatrix}.$$

Gluon stress tensor in QCD Lagrangian has the form

$$F^\alpha_{\mu\nu} = \partial_\mu A^\alpha_\nu - \partial_\nu A^\alpha_\mu + g_s f^{\alpha\beta\gamma} A^\beta_\mu A^\gamma_\nu, \qquad (5.44)$$

with the nonzero structure constant antisymmetric on all indices of the gauge group:

$$f^{123} = 1, \quad f^{147} = f^{246} = f^{257} = f^{345} = \frac{1}{2},$$

$$f^{156} = f^{367} = -\frac{1}{2}, \quad f^{458} = f^{678} = \frac{\sqrt{3}}{2}, \qquad (5.45)$$

where $[t^\alpha, t^\beta] = if^{\alpha\beta\gamma}t^\gamma$, $\alpha, \beta, \gamma = 1, \dots, 8$. The mass terms $-m_q \bar{q}^i q_i$ are not included as far as they are present in the electroweak Lagrangian (5.17).

The choice of Gell-Mann matrices in the form (5.43) fix the basis in $SU(3)$. This enables us to write out the covariant derivatives (5.42) in the explicit form

$$D_\mu = \mathbf{I}\partial_\mu - i\frac{g_s}{2}\begin{pmatrix} A^3_\mu + \dfrac{1}{\sqrt{3}}A^8_\mu & A^1_\mu - iA^2_\mu & A^4_\mu - iA^5_\mu \\ A^1_\mu + iA^2_\mu & \dfrac{1}{\sqrt{3}}A^8_\mu - A^3_\mu & A^6_\mu - iA^7_\mu \\ A^4_\mu + iA^5_\mu & A^6_\mu + iA^7_\mu & -\dfrac{2}{\sqrt{3}}A^8_\mu \end{pmatrix}$$

$$= \mathbf{I}\partial_\mu - i\frac{g_s}{2}\begin{pmatrix} A^{RR}_\mu & A^{RG}_\mu & A^{RB}_\mu \\ A^{GR}_\mu & A^{GG}_\mu & A^{GB}_\mu \\ A^{BR}_\mu & A^{BG}_\mu & A^{BB}_\mu \end{pmatrix}, \qquad (5.46)$$

where

$$A_\mu^{RR} = \frac{1}{\sqrt{3}} A_\mu^8 + A_\mu^3, \quad A_\mu^{GG} = \frac{1}{\sqrt{3}} A_\mu^8 - A_\mu^3, \quad A_\mu^{BB} = -\frac{2}{\sqrt{3}} A_\mu^8,$$

$$A_\mu^{RR} + A_\mu^{GG} + A_\mu^{BB} = 0, \quad A_\mu^{GR} = A_\mu^1 + iA_\mu^2 = \bar{A}_\mu^{RG}, \tag{5.47}$$

$$A_\mu^{BR} = A_\mu^4 + iA_\mu^5 = \bar{A}_\mu^{RB}, \quad A_\mu^{BG} = A_\mu^6 + iA_\mu^7 = \bar{A}_\mu^{GB},$$

and Lagrangian (5.41)

$$\mathcal{L} = \bar{u}_i(i\gamma^\mu)(D_\mu)^{ij} u_j + \cdots - \frac{1}{4} F_{\mu\nu}^\alpha F^{\mu\nu\,\alpha}$$

$$\equiv L_u + \cdots - \frac{1}{4} F_{\mu\nu}^\alpha F^{\mu\nu\,\alpha}, \tag{5.48}$$

where only the u-quark part is given. Let us note that, in QCD, the special mechanism of spontaneous symmetry breaking, which provides masses for vector bosons, is absent; therefore, gluons are massless particles.

The QCD Lagrangian has rich dynamical content. It describes complicated hadron spectrum, color confinement, asymptotic freedom and many other effects.

5.5 QCD with contracted gauge group

The contracted special unitary group $SU(3;\kappa)$ is defined by the action

$$q'(\kappa) = \begin{pmatrix} q_1' \\ \kappa_1 q_2' \\ \kappa_1\kappa_2 q_3' \end{pmatrix} = \begin{pmatrix} u_{11} & \kappa_1 u_{12} & \kappa_1\kappa_2 u_{13} \\ \kappa_1 u_{21} & u_{22} & \kappa_2 u_{23} \\ \kappa_1\kappa_2 u_{31} & \kappa_2 u_{32} & u_{33} \end{pmatrix} \begin{pmatrix} q_1 \\ \kappa_1 q_2 \\ \kappa_1\kappa_2 q_3 \end{pmatrix}$$

$$= U(\kappa)q(\kappa), \quad \det U(\kappa) = 1, \quad U(\kappa)U^\dagger(\kappa) = 1 \tag{5.49}$$

on the complex space $\mathbf{C}_3(\kappa)$ in such a way that the hermitian form

$$q^\dagger(\kappa)q(\kappa) = |q_1|^2 + \kappa_1^2(|q_2|^2 + \kappa_2^2|q_3|^2) \tag{5.50}$$

remains invariant, when the contraction parameters tend to zero: $\kappa_1, \kappa_2 \to 0$. Transition from the classical group $SU(3)$ and space $\mathbf{C}_3$ to the group $SU(3;\kappa)$ and space $\mathbf{C}_3(\kappa)$ is given by the substitution

$$q_1 \to q_1, \quad q_2 \to \kappa_1 q_2, \quad q_3 \to \kappa_1\kappa_2 q_3,$$

$$A_\mu^{GR} \to \kappa_1 A_\mu^{GR}, \quad A_\mu^{BG} \to \kappa_2 A_\mu^{BG}, \quad A_\mu^{BR} \to \kappa_1\kappa_2 A_\mu^{BR}, \tag{5.51}$$

and diagonal gauge fields $A_\mu^{RR}, A_\mu^{GG}, A_\mu^{BB}$ remain unchanged.

Substituting (5.51) in (5.48), we obtain the quark part of the Lagrangian in the form

$$
\mathcal{L}_q(\kappa) = \sum_q \left\{ i\bar{q}_1 \gamma^\mu \partial_\mu q_1 + \frac{g_s}{2} |q_1|^2 \gamma^\mu A_\mu^{RR} \right.
$$

$$
+ \kappa_1^2 \left[i\bar{q}_2 \gamma^\mu \partial_\mu q_2 + \frac{g_s}{2} \left(|q_2|^2 \gamma^\mu A_\mu^{GG} + q_1 \bar{q}_2 \gamma^\mu A_\mu^{GR} + \bar{q}_1 q_2 \gamma^\mu \bar{A}_\mu^{GR} \right) \right]
$$

$$
+ \kappa_1^2 \kappa_2^2 \left[i\bar{q}_3 \gamma^\mu \partial_\mu q_3 + \frac{g_s}{2} \left(|q_3|^2 \gamma^\mu A_\mu^{BB} + q_1 \bar{q}_3 \gamma^\mu A_\mu^{BR} + \bar{q}_1 q_3 \gamma^\mu \bar{A}_\mu^{BR} \right. \right.
$$

$$
\left. \left. \left. + q_2 \bar{q}_3 \gamma^\mu A_\mu^{BG} + \bar{q}_2 q_3 \gamma^\mu \bar{A}_\mu^{BG} \right) \right] \right\} = L_q^\infty + \kappa_1^2 L_q^{(2)} + \kappa_1^2 \kappa_2^2 L_q^{(4)}.
$$

$$(5.52)$$

Let us introduce the notations

$$
\partial A^k \equiv \partial_\mu A_\nu^k - \partial_\nu A_\mu^k, \quad [k, m] \equiv A_\mu^k A_\nu^m - A_\mu^m A_\nu^k; \qquad (5.53)
$$

then, the gluon tensor (5.44) has the following components

$$
F_{\mu\nu}^1 = \kappa_1 \left\{ \partial A^1 + \frac{g_s}{2} (2[2,3] + \kappa_2^2([4,7] - [5,6])) \right\},
$$

$$
F_{\mu\nu}^2 = \kappa_1 \left\{ \partial A^2 + \frac{g_s}{2} (-2[1,3] + \kappa_2^2([4,6] + [5,7])) \right\},
$$

$$
F_{\mu\nu}^3 = \partial A^3 + \frac{g_s}{2} (\kappa_1^2 2[1,2] - \kappa_2^2[6,7] + \kappa_1^2 \kappa_2^2[4,5]),
$$

$$
F_{\mu\nu}^4 = \kappa_1 \kappa_2 \left\{ \partial A^4 - \frac{g_s}{2} ([1,7] + [2,6] + [3,5] - \sqrt{3}[5,8]) \right\},
$$

$$
F_{\mu\nu}^5 = \kappa_1 \kappa_2 \left\{ \partial A^5 + \frac{g_s}{2} ([1,6] - [2,7] + [3,4] - \sqrt{3}[4,8]) \right\},
$$

$$(5.54)$$

$$
F_{\mu\nu}^6 = \kappa_2 \left\{ \partial A^6 + \frac{g_s}{2} (\kappa_1^2([2,4] - [1,5]) + [3,7] + \sqrt{3}[7,8]) \right\},
$$

$$
F_{\mu\nu}^7 = \kappa_2 \left\{ \partial A^7 + \frac{g_s}{2} (\kappa_1^2([1,4] + [2,5]) - [3,6] - \sqrt{3}[6,8]) \right\},
$$

$$
F_{\mu\nu}^8 = \partial A^8 + \frac{g_s \sqrt{3}}{2} \kappa_2^2 (\kappa_1^2[4,5] + [6,7]).
$$

Gluon part of Lagrangian (5.48) is as follows

$$\mathcal{L}_{gl}(\kappa) = -\frac{1}{4}F^{\alpha}_{\mu\nu}F^{\mu\nu\,\alpha}$$

$$= -\frac{1}{4}\{H_3^2 + H_8^2 + \kappa_1^2(F_1^2 + F_2^2 + 2H_3F_3)$$

$$+ \kappa_2^2(G_6^2 + G_7^2 + 2H_3G_3 - 2\sqrt{3}H_8G_3) + \kappa_1^4F_3^2 + \kappa_2^4 4G_3^2$$

$$+ \kappa_1^2\kappa_2^2[P_4^2 + P_5^2 + 2(F_1G_1 + F_2G_2 + F_3G_3 + F_6G_6 + F_7G_7$$

$$+ \sqrt{3}H_8P_3)] + \kappa_1^2\kappa_2^4(G_1^2 + G_2^2 - 4G_3P_3)$$

$$+ \kappa_1^4\kappa_2^2(F_6^2 + F_7^2 + 2F_3P_3) + \kappa_1^4\kappa_2^4 4P_3^2\}, \tag{5.55}$$

where

$$F_1 = \partial A^1 + g_s[2,3], \quad F_2 = \partial A^2 - g_s[1,3],$$

$$G_1 = \frac{g_s}{2}([4,7] - [5,6]), \quad G_2 = \frac{g_s}{2}([4,6] + [5,7]),$$

$$H_3 = \partial A^3, \quad F_3 = g_s[1,2], \quad G_3 = -\frac{g_s}{2}[6,7], \quad P_3 = \frac{g_s}{2}[4,5],$$

$$P_4 = \partial A^4 - \frac{g_s}{2}([1,7] + [2,6] + [3,5] - \sqrt{3}[5,8]),$$

$$P_5 = \partial A^5 + \frac{g_s}{2}([1,6] - [2,7] + [3,4] - \sqrt{3}[4,8]), \tag{5.56}$$

$$G_6 = \partial A^6 + \frac{g_s}{2}([3,7] + \sqrt{3}[7,8]), \quad F_6 = \frac{g_s}{2}([2,4] - [1,5]),$$

$$G_7 = \partial A^7 - \frac{g_s}{2}([3,6] + \sqrt{3}[6,8]), \quad F_7 = \frac{g_s}{2}([1,4] + [2,5]),$$

$$H_8 = \partial A^8.$$

In the framework of Cayley–Klein scheme the gauge group $SU(3;\kappa)$ has two one-parameter contractions $\kappa_1 \to 0, \kappa_2 = 1$ and $\kappa_2 \to 0, \kappa_1 = 1$, as well as one two-parameter contraction $\kappa_1, \kappa_2 \to 0$. We consider the following contraction: $\kappa_1 = \kappa_2 = \kappa = \epsilon \to 0$ and the corresponding limit of QCD. The simple algebra $su(3)$ obtains the structure of a semidirect sum after this contraction

$$su(3;0) = T_4 \Subset ((T_2 \Subset u(1)) \oplus \hat{u}(1))$$

$$= (T_4 \Subset T_2) \Subset (u(1) \oplus \hat{u}(1)), \tag{5.57}$$

where $T_4 = \{t^1, t^2, t^4, t^5\}$, $T_2 = \{t^6, t^7\}$ are Abelian subalgebras, $\hat{u}(1) = \{\frac{1}{\sqrt{3}} t^8 - t^3\}$ and $u(1) = \{\frac{1}{\sqrt{3}} t^8 + t^3\}$ are one-dimensional subalgebras. The simple group $SU(3)$ obtains the structure of semidirect product.

We take the contraction parameter κ as identical to the contraction parameter ϵ (5.22) of the Electroweak Model: $\kappa = \epsilon$, so the limit $\kappa \to 0$ corresponds to the infinite temperature limit. The quark part of Lagrangian (5.52) is represented as a sum of terms proportional to zero, the second and forth powers of the contraction parameter

$$\mathcal{L}_q(\epsilon) = L_q^\infty + \epsilon^2 L_q^{(2)} + \epsilon^4 L_q^{(4)}, \tag{5.58}$$

and gluon part (5.55) is represented as a sum

$$\mathcal{L}_{gl}(\epsilon) = L_{gl}^\infty + \epsilon^2 L_{gl}^{(2)} + \epsilon^4 L_{gl}^{(4)} + \epsilon^6 L_{gl}^{(6)} + \epsilon^8 L_{gl}^{(8)}, \tag{5.59}$$

where

$$L_{gl}^\infty = -\frac{1}{4}\{(\partial A^3)^2 + (\partial A^8)^2\},$$

$$L_{gl}^{(2)} = -\frac{1}{4}\left\{ (\partial A^1 + g_s[2,3])^2 + \left(\partial A^6 + \frac{g_s}{2}([3,7] + \sqrt{3}[7,8]) \right)^2 \right.$$

$$+ (\partial A^2 - g_s[1,3])^2 + \left(\partial A^7 - \frac{g_s}{2}([3,6] + \sqrt{3}[6,8]) \right)^2$$

$$\left. + g_s((2[1,2] - [6,7])\partial A^3 + \sqrt{3}[6,7]\partial A^8) \right\}$$

$$L_{gl}^{(4)} = -\frac{1}{4}\left\{ (\partial A^4)^2 + (\partial A^5)^2 + g_s(([4,7] - [5,6])\partial A^1 \right.$$

$$+ ([4,6] + [5,7])\partial A^2 - ([1,7] + [2,6] + [3,5] - \sqrt{3}[5,8])\partial A^4$$

$$+ ([1,6] - [2,7] + [3,4] - \sqrt{3}[4,8])\partial A^5 + ([2,4] - [1,5])\partial A^6$$

$$+ ([1,4] + [2,5])\partial A^7 + \sqrt{3}[4,5]\partial A^8) + g_s^2 \left([1,2]^2 + [6,7]^2 \right.$$

$$- [1,2][6,7] - [1,3]([4,6] + [5,7]) + [2,3]([4,7] - [5,6])$$

$$+ \frac{1}{2}([3,7] + \sqrt{3}[7,8])([2,4] - [1,5])$$

$$- \frac{1}{2}([3,6] + \sqrt{3}[6,8])([1,4] + [2,5])$$

$$+ \frac{1}{2}([1,7] + [2,6] + [3,5] - \sqrt{3}[5,8])^2$$

$$+ \frac{1}{2}([1,6] - [2,7] + [3,4] - \sqrt{3}[4,8])^2 \bigg) \bigg\},$$

$$L_{gl}^{(6)} = -\frac{g_s^2}{16} \{([4,7] - [5,6])^2 + ([4,6] + [5,7])^2$$

$$+ ([2,4] - [1,5])^2 + ([1,4] + [2,5])^2 + 4([1,2] + [6,7])[4,5]\},$$

$$L_{gl}^{(8)} = -\frac{g_s^2}{4} [4,5]^2. \tag{5.60}$$

In the infinite energy (temperature) limit $\kappa = \epsilon \to 0$, most parts of gluon tensor components are equal to zero and the expressions for two nonzero components are simplified as

$$F_{\mu\nu}^3 = \partial_\mu A_\nu^3 - \partial_\nu A_\mu^3 = \frac{1}{2} \left(F_{\mu\nu}^{RR} - F_{\mu\nu}^{GG} \right),$$

$$F_{\mu\nu}^8 = \partial_\mu A_\nu^8 - \partial_\nu A_\mu^8 = \frac{\sqrt{3}}{2} \left(F_{\mu\nu}^{RR} + F_{\mu\nu}^{GG} \right), \tag{5.61}$$

so the QCD Lagrangian $\mathcal{L}_\infty$ in this limit can be explicitly expressed as

$$\mathcal{L}_\infty = L_q^\infty + L_{gl}^\infty = \sum_q \left\{ i\bar{q}_R \gamma^\mu \partial_\mu q_R + \frac{g_s}{2} |q_R|^2 \gamma^\mu A_\mu^{RR} \right\}$$

$$- \frac{1}{4} \left(F_{\mu\nu}^{RR} \right)^2 - \frac{1}{4} \left(F_{\mu\nu}^{GG} \right)^2 - \frac{1}{4} F_{\mu\nu}^{RR} F_{\mu\nu}^{GG}. \tag{5.62}$$

From $\mathcal{L}_\infty$ we conclude that only the dynamic terms for the first color component of massless quarks survive under the infinite temperature, which means that the quarks are monochromatic, and the surviving terms describe the interactions of these components with R-gluons. Besides R-gluons there are also G-gluons, which do not interact with the quarks.

Similar to the Electroweak Model starting with $\mathcal{L}_q(\epsilon)$ (5.58) and $\mathcal{L}_{gl}(\epsilon)$ (5.59) one can construct a number of intermediate models for QCD, which describe the restoration of color degrees of freedom for the quarks and the gluon interactions in the Universe's evolution.

It follows from Lagrangian $\mathcal{L}(\epsilon) = \mathcal{L}_q(\epsilon) + \mathcal{L}_{gl}(\epsilon)$, that the total reconstruction of the quark color degrees of freedom will take place after the restoration of all quark masses ($\approx \epsilon^2$) at the same time with the reestablishment of all electroweak interactions ($\approx \epsilon^4$). All color interactions start to work later because some of them are proportionate to the eighth power of the contraction parameter ϵ^8.

5.6 Estimation of boundary values

As it was mentioned the contraction of the gauge group of QCD gives an opportunity to arrange in time different stages of its development, but does not make it possible to identify their absolute date. Let us try to estimate this date with the help of additional assumptions. The equality of the contraction parameters for QCD and the EWM is one of these assumptions.

Then we use the fact (see Fig. 5.1) that the electroweak epoch starts at the temperature $E_4 = 100\,\text{GeV}$ and the QCD epoch begins at $E_8 = 0.2\,\text{GeV}$. In other words, we assume that complete reconstruction of the EWM, whose Lagrangian has minimal terms proportionate to ϵ^4, and QCD, whose Lagrangian has minimal terms proportionate to ϵ^8, take place at these temperatures. Let us denote by Δ the cutoff level for ϵ^k, $k = 1, 2, 4, 6, 8$, i.e. for $\epsilon^k < \Delta$ all the terms proportionate to ϵ^k are negligible quantities in the Lagrangian. At last, we suppose that the contraction parameter inversely depends on temperature

$$\epsilon(T) = \frac{A}{T}, \tag{5.63}$$

where A is constant.

As far as the minimal terms in the QCD Lagrangian are proportional to ϵ^8 and QCD is completely reconstructed at $T_8 = 0.2\,\text{GeV}$, we have the equation $\epsilon^8(T_8) = A^8 T_8^{-8} = \Delta$ and obtain $A = T_8 \Delta^{1/8} = 0.2\Delta^{1/8}\,\text{GeV}$. The minimal terms in the EWM Lagrangian are proportional to ϵ^4 and it is reconstructed at $T_4 = 100\,\text{GeV}$, so we have $\epsilon^4(T_4) = A^4 T_4^{-4} = \Delta$, i.e. $T_4 = A\Delta^{-1/4} = T_8 \Delta^{1/8}\Delta^{-1/4} = T_8\Delta^{-1/8}$ and we obtain the cutoff level $\Delta = (T_8 T_4^{-1})^8 = (0, 2 \cdot 10^{-2})^8 \approx 10^{-22}$, which is consistent with the typical temperatures of the Standard Model. From the equation $\epsilon^k(T_k) = A^k T_k^{-k} = \Delta$ we obtain

$$T_k = \frac{A}{\Delta^{1/k}} = \frac{T_8 \Delta^{1/8}}{\Delta^{1/k}} = T_8 \Delta^{\frac{k-8}{8k}} \approx 10^{\frac{88-15k}{4k}}\,\text{GeV}. \tag{5.64}$$

Simple calculations give the following estimations for the boundary values of the average energy (or temperature) in the early Universe (GeV): $T_1 = 10^{18}$, $T_2 = 10^7$, $T_3 = 10^3$, $T_4 = 10^2$, $T_6 = 1$, $T_8 = 2 \cdot 10^{-1}$. The obtained estimation for the "infinity" energy $T_1 \approx 10^{18}\,\text{GeV}$ is comparable with Planck energy $\approx 10^{19}\,\text{GeV}$, where the gravitation effects are important. So the developed evolution of the elementary particles does not exceed the range of the problems described by the electroweak and strong interactions.

It should be noted that for the power function class $\epsilon(T) = BT^{-p}$ the estimations for energy boundary values are very weakly dependent on power p. So practically we obtain the same T_k for $p = 10$ as for the simplest function (5.63) with $p = 1$.

5.7 Concluding remarks

We have investigated the high-temperature limit of the Standard Model which was obtained from the first principles of the gauge theory as the contraction of its gauge group. It was shown that the mathematical contraction parameter is inversely proportional to the temperature and its zero limit corresponds to the infinite temperature limit of the SM. In this limit the Standard Model passes through several stages, distinguished by the powers of the contraction parameter, which gives the opportunity to classify them in time as earlier-later. To determine the absolute date of these stages the additional assumptions were used, namely: the inverse dependence of ϵ on the temperature (5.63) and the cutoff level Δ for ϵ^k. The unknown parameters are determined with the help of the QCD and EWM typical energies.

The exact expressions for the respective Lagrangians for any stage in the Standard Model evolution are obtained. With the help of decompositions (5.37), (5.58), (5.59) the intermediate models $\mathcal{L}_k$ for any temperature scale are constructed. It gives an opportunity to draw conclusions on the interactions and properties of the elementary particles in each of the considered epochs. The presence of several intermediate models, instead of a single, in the interval between Planck energy 10^{19} GeV and the EWM typical energy 10^2 GeV automatically takes away the so-called hierarchy problem of the Standard Model [Emel'yanov (2007)].

At the infinite temperature limit ($T > 10^{18}$ GeV) all particles including vector bosons lose their masses and the electroweak interactions become long-range. Monochromatic massless quarks are exchanged by only one sort of R-gluons. Besides R-gluons there are also G-gluons, which do not interact with quarks. It follows from the explicit form of Lagrangians $L_\infty^{int}(A_\mu, Z_\mu)$ (5.39) and $\mathcal{L}_\infty$ (5.62) that only the particles of the same sort interact with each other. Particles of different sorts do not interact. It looks like some stratification of leptons and quark-gluon plasma with only one sort of particles in each stratum.

At the level of classical gauge fields it is already possible to give some conclusions on the appearance of the elementary particles mass on the different stages of the Universe's evolution. In particular we can conclude that half of quarks ($\approx\epsilon$, $10^{18}\,\text{GeV} > T > 10^{7}\,\text{GeV}$) first restore their mass. Then Z-bosons, electrons and other quarks become massive ($\approx\epsilon^2$, $10^{7}\,\text{GeV} > T > 10^{3}\,\text{GeV}$). Finally Higgs boson χ and charged $W^{\pm}$-bosons restore their masses because their mass terms are multiplied by ϵ^4 ($T < 10^2\,\text{GeV}$).

In a similar way it is possible to describe the evolution of particle interactions. Self-action of Higgs boson appears with its mass restoration. At the same epoch, interactions of four $W^{\pm}$-bosons, as well as of two Higgs and two W-bosons (5.31) begin. The only one term in the Lagrangian, which is proportional to the third power of ϵ, describes the interaction of Higgs boson with charged $W^{\pm}$-bosons ($T < 10^3\,\text{GeV}$). The rest of the electroweak particle interactions appear in the second order of the contraction parameter ($10^{7}\,\text{GeV} > T > 10^{3}\,\text{GeV}$). Some part of color interactions between quarks in Lagrangian (5.52) is proportional to ϵ^2 ($T < 10^{7}\,\text{GeV}$) and the rest is proportional to ϵ^4 ($T < 10^2\,\text{GeV}$). Therefore the complete restoration of quark color degrees of freedom takes place after the appearance of quark masses ($\approx\epsilon^2$, $T < 10^{7}\,\text{GeV}$) (5.40) together with the restoration of all electroweak interactions ($\approx\epsilon^4$, $T < 10^2\,\text{GeV}$). Complete color interactions start later because they are proportional to ϵ^8 ($T < 10^{-1}\,\text{GeV}$).

The evolution of the elementary particles and their interactions in the early Universe obtained with the help of the contractions of the gauge groups of the Standard Model does not contradict the canonical one (see [Gorbunov and Rubakov (2011)] and Fig. 5.1), according to which the QCD phase transitions take place later than the electroweak phase transitions. The developed evolution of the Standard Model presents the basis for a more detailed analysis of different phases in the formation of leptons and quark-gluon plasma, considering that the terms $L_{gl}^{(6)}$ and $L_{gl}^{(8)}$ (5.60) in the gluon Lagrangian (5.59) of QCD become negligible at temperatures from $0.2\,\text{GeV}$ to $100\,\text{GeV}$.

Bibliography

Abellanas, L. and Martinez Alonso L. (1975). A general setting for Casimir invariants, *J. Math. Phys.* **16**, 8, pp. 1580–1584.

Bacry, H. and Levy-Leblond, J.-M. (1968). Possible kinematics, *J. Math. Phys.* **9**, 10, pp. 1605–1614.

Bacry, H. and Nuyts J. (1986). Classification of ten-dimensional kinematical groups with space isotropy, *J. Math. Phys.* **27**, 10, pp. 2455–2457.

Ballesteros, A., Gromov, N. A., Herranz, F. J., del Olmo, M. A. and Santander, M. (1995). Lie bialgebra contractions and quantum deformations of quasi-orthogonal algebras, *J. Math. Phys.* **36**, pp. 5916–5936.

Bargmann, V. (1954). On unitary ray representations of continuous groups, *Ann. Math.* **59**, 1, pp. 1–46.

Barut, A. O. and Raczka, R. (1977). *Theory of Group Representations and Applications* (PWN—Polish Scientific Publishers, Warszawa).

Bateman, H. and Erdelyi, A. (1953). *Higher Transcendental Functions* (Mc Graw–Hill, New York, Toronto, London).

Beresin, A. V., Kurochkin, Yu. A. and Tolkachev, E. A. (1989). *Quaternions in Relativistic Physics* (Nauka i tehnika, Minsk) (in Russian).

Beresin, F. A. (1966). *The Method of Second Quantization, Monographs and Textbooks in Pure and Applied Physics*, Vol. 24 (Academic Press, New York, London).

Beresin, F. A. (1987). *Introduction to Superanalysis* (Springer, Berlin, Heidelberg, New York).

Biedenharn, L. C. and Louck, J. (1981). *Angular Momentum in Quantum Physics* (Addison-Wesley, London).

Blokh, A. Sh. (1982). *Numerical Systems* (Vyssheishaya shkola, Minsk) (in Russian).

Bourbaki, N. (2007). *Varietes Differentielles et Analytiques: Fascicule de Resultats* (Springer, Berlin, Heidelberg, New York).

Celeghini, E. and Tarlini, M. (1981). Contractions of group representations. I, *Nuovo Cimento* **B61**, 2, pp. 265–277.

Celeghini, E. and Tarlini, M. (1981). Contractions of group representations. II, *Nuovo Cimento* **B65**, 1, pp. 172–180.

Celeghini, E. and Tarlini M. (1982). Contractions of group representations. III, *Nuovo Cimento* **B68**, 1, pp. 133–141.

Celeghini, E., Giachetti, R., Sorace, E. and Tarlini M. (1990). Three-dimensional quantum groups from contractions of $SU_q(2)$, *J. Math. Phys.* **31**, pp. 2548–2551.

Celeghini, E., Giachetti, R., Kulish, P. P., Sorace, E. and Tarlini M. (1991). Hopf superalgebra contractions and R-matrix for fermions, *J. Phys. A: Math. Gen.* **24**, 24, pp. 5675–5682.

Celeghini, E., Giachetti, R., Sorace, E. and Tarlini M. (1991). The three-dimensional Euclidean quantum group $E(3)_q$ and its R-matrix, *J. Math. Phys.* **32**, 5, pp. 1159–1165.

Celeghini, E., Giachetti, R., Sorace, E. and Tarlini M. (1991). The quantum Heisenberg qroup $H(1)_q$, *J. Math. Phys.* **32**, 5, pp. 1155–1158.

Celeghini, E., Giachetti, R., Sorace, E. and Tarlini M. (1992). Contractions of quantum groups, in *Quantum Groups* (Springer, Berlin), pp. 21–244. (Lecture Notes in Mathematics, **1510**).

Chaichian, M., Demichev, A. P. and Nelipa, N. F. (1983). The Casimir operators of inhomogeneous groups, *Commun. Math. Phys.* **90**, pp. 353–372.

Chakrabarti, A. (1968). Class of representations of the $IU(n)$ and $IO(n)$ algebras and respective deformations to $U(n,1), O(n,1)$, *J. Math. Phys.* **9**, 12, pp. 2087–2100.

Clifford, W. K. (1873). Preliminary sketch of biquaternions, *Proc. London Math. Soc.* **4**, pp. 381–395.

Dattoli, G., Richetta, M. and Torre, A. (1988). Evolution of $SU(2)$ and $SU(1,1)$ states: A further mathematical analysis, *J. Math. Phys.* **29**, pp. 2586–2592.

de Prunele, E. (1988). $SU(1,1)$, its connections with $SU(2)$, and the vector model, *J. Math. Phys.* **29**, pp. 2523–2528.

Derom, J.-R. and Dubois, J.-G. (1972). Hooke's symmetries and nonrelativistic cosmological kinematics, *Nuovo Cimento* **9B**, pp. 351–376.

Dodonov, V. V. and Man'ko, V. I. (1987). *P. N. Lebedev Physical Institute Proceedings* **183**, pp. 182–288 (in Russian).

Dubois, J.-G. (1973). Hooke's symmetries and nonrelativistic cosmological kinematics. II. Irreducible projective representations, *Nuovo Cimento* **15B**, 1, pp. 1–24.

Duplii, S. A. (1988). Nilpotent mechanics and supersymmetry, *Probl. Nucl. Phys. Cosmic Rays* **30**, pp. 41–48 (in Russian).

Emel'yanov, V. M. (2007). *Standard Model and its Expansion* (Fizmatlit, Moscow) (in Russian).

Fedorov, F. I. (1979). *Lorentz Group* (Nauka, Moscow) (in Russian).

Fernandez Sanjuan, M. A. (1984). Group contraction and nine Cayley–Klein geometries, *Int. J. Theor. Phys.* **23**, 1, pp. 1–14.

Gel'fand, I. M. (1950). Centre of infinitesimal group ring, *Math. Collection* **26(28)**, 1, pp. 103–112 (in Russian).

Gel'fand, I. M. and Graev, M. I. (1965). Finite dimensional irreducible representations of unitary and general linear groups and connected special functions, *Proc. AS USSR. Math. Ser.* **29**, pp. 1329–1356 (in Russian).

Gel'fand, I. M. and Graev, M. I. (1967). Irreducible representations of Lie algebra of $U(p, q)$ group, in *Higher Energy Physics and Theory of Elementary Particles* (Kiev, Naukova dumka), pp. 216–226. (in Russian).

Gel'fand, I. M., Minlos, R. A. and Shapiro, Z. Ya. (1963). *Representations of the Rotation and Lorentz Groups and their Applications* (Pergamon Press).

Gel'fand, I. M. and Tsetlin, M. L. (1950). Finite-dimensional representations of unimodular matrix group, *Soviet Math. Dokl.* **71**, 5, pp. 825–828 (in Russian).

Gershun, V. D. and Tkach, V. I. (1984). Para-Grassmann variables and description of massive particles with unit spin, *Ukrainian Phys. J.* **29**, 11, pp. 1620–1627 (in Russian).

Gershun, V. D. and Tkach, V. I. (1985). Grassmann and para-Grassmann variables and dynamics of massless particles with unit spin, *Probl. Nucl. Phys. Cosmic Rays* **23**, pp. 42–60 (in Russian).

Glauber, R. J. (1963). Coherent and incoherent states of the radiation field, *Phys. Rev.* **131**, pp. 2766–2788.

Gorbunov, D. S. and Rubakov, V. A. (2011). *Introduction to the Theory of the Early Universe: Hot Big Bang Theory* (World Scientific, Singapore, 2011).

Gromov, N. A. (1981). Casimir operators of groups of motions of space of constant curvature, *Teor. Mat. Fiz.* **49**, 2, pp. 210–218 (in Russian).

Gromov, N. A. (1984). Analogs of F. I. Fedorov's parametrization of groups $SO_3(j)$, $SO_4(j)$ in fiber spaces, *Vesci AS BSSR. Ser. fiz.-math.* **2**, pp. 108–114 (in Russian).

Gromov, N. A. (1984) *Special Unitary Groups in Fiber Spaces* (Syktyvkar) (in Russian).

Gromov, N. A. (1990). *Contractions and Analytical Continuations of Classical Groups. Unified Approach* (Komi Science Centre RAS, Syktyvkar) (in Russian).

Gromov, N. A. (1990). Transitions: contractions and analytical continuations of the Cayley–Klein groups, *Int. J. Theor. Phys.* **29**, 6, pp. 607–620.

Gromov, N. A. and Man'ko, V. I. (1990). The Jordan–Schwinger representations of Cayley–Klein groups. *J. Math. Phys.* **31**, 5, I. The orthogonal groups, pp. 1047–1053, II. The unitary groups, pp. 1054–1059; III. The symplectic groups, pp. 1060–1064.

Gromov, N. A. and Man'ko, V. I. (1991). Group quantum systems in correlated coherent states representations, *P. N. Lebedev Physical Institute Proceedings* **200**, pp. 3–55 (in Russian).

Gromov, N. A. (1991). The Gel'fand–Tsetlin representations of the unitary Cayley–Klein algebras, *J. Math. Phys.* **32**, 4, pp. 837–844.

Gromov, N. A. (1992). The Gel'fand–Tsetlin representations of the orthogonal Cayley–Klein algebras, *J. Math. Phys.* **33**, 4, pp. 1363–1373.

Gromov, N. A. (2004). From Wigner–Inönü group contraction to contractions of algebraical structures, *Acta Physica Hungarica. Sec. A. Heavy Ion Physics* **19**, 3–4, pp. 209–212.

Gromov, N. A. (2012). *Contractions of Classical and Quantum Groups* (Fizmatlit, Moscow) (in Russian).

Gromov, N. A. (2015). Natural limits of electroweak model as contraction of its gauge group, *Physica Scripta* **90**, 074009 (9 pp).

Gromov, N. A. (2016). Elementary particles in the early Universe, *J. Cosmol. Astroparticle Phys.* **03**, 053 (15 pp).

Inönü, E. and Wigner, E. P. (1953). On the contraction of groups and their representations, *Proc. Nat. Acad. Sci. USA* **39**, pp. 510–524.

Jordan, P. (1935). Der Zusammenhang der Symmetrischen und Linearen Gruppen und das Mehrkorperproblem, *Z. Phys.* **94**, pp. 531–535.

Kim, Y. S. and Wigner, E. P. (1987). Cylindrical group and massless particles, *J. Math. Phys.* **28**, 5, pp. 1175–1179.

Kim, S. K. (1987). Theorems on Jordan–Schwinger representations of Lie algebras, *J. Math. Phys.* **28**, pp. 2540–2545.

Kisil V. V. (2012). *Geometry of Mobius Transformations: Elliptic, Parabolic and Hyperbolic Actions of* $SL_2(\mathbf{R})$ (Imperial College Press, London).

Kisil V. V. (2012). Is commutativity of observables the main feature, which separate classical mechanics from quantum, *Proc. Komi Science Centre RAS* **3(11)**, pp. 4–9 (in Russian).

Kisil V. V. (2012). *Classical/Quantum = Commutative/Noncommutative*, arXiv: 1204.1858v2.

Konopleva, N. P. and Popov, V. N. (1981). *Gauge Fields* (Harwood, Chur, London, New York).

Korn, G. A. and Korn, T. M. (1961). *Mathematical Handbook* (McGraw-Hill, New York, 1961).

Kotel'nikov, A. P. (1895). *Screw Calculus and Some Applications in Geometry and Mechanics* (Kazan) (in Russian).

Kuriyan, J. G., Mukunda, N. and Sudarshan E. C. G. (1968). Master analytic representations: reduction of $O(2,1)$ in an $O(1,1)$ basis, *J. Math. Phys.* **9**, 12, pp. 2100–2108.

Kuriyan, J. G., Mukunda, N. and Sudarshan E. C. G. (1968). Master analytic representations and unified representation theory of certain orthogonal and pseudo-orthogonal groups, *Comm. Math. Phys.* **8**, 3, pp. 204–227.

Levy-Leblond, J.-M. (1965). Une nouvelle limite non-relativiste du groupe de Poincaré, *Ann. Inst. H. Poincaré* **A3**, 1, pp. 1–12.

Leznov, A. N., Malkin, I. A. and Man'ko, V. I. (1977). Canonical transformations and representation theory of Lie groups, *Proc. Lebedev Inst. RAS* **96**, pp. 24–71 (in Russian).

Linblad, G. and Nagel, B. (1970). Continuous bases for unitary irreducible representations of $SU(1,1)$, *Ann. Inst. H. Poincaré* **13**, 1, pp. 27–56.

Linde, L. D. (1990). *Particle Physics and Inflationary Cosmology* (Nauka, Moscow) (in Russian).

Lohe, M. A. and Hurst, C. A. (1971). *J. Math. Phys.* **12**, pp. 1882.

Lord, E. A. (1985). Geometrical interpretation of Inönü–Wigner contractions, *Int. J. Theor. Phys.* **24**, 7, pp. 723–730.

Lyhmus, Ya. H. (1969). *Limited (Contracted) Lie Groups* (Tartu) (in Russian).

Malkin, I. A. and Man'ko, V. I. (1979). *Dynamical Symmetries and Coherent States of Quantum Systems* (Nauka, Moscow) (in Russian).

Man'ko, V. I. and Trifonov, D. A. (1987). In *Group Theoretical Methods in Physics* (Harwood Academic, London), Vol. 3, p. 795–805.

Montigny, M. de and Patera, J. (1991). Discrete and continuous graded contractions of Lie algebras and superalgebras, *J. Phys. A: Math. Gen.* **24**, pp. 525–547.

Montigny, M. de (1994). Graded contractions of bilinear invariant forms of Lie algebra, *J. Phys. A: Math. Gen.* **27**, 13, pp. 4537–4548.

Montigny, M. de (1996). Graded contractions of affine Lie algebras, *J. Phys. A: Math. Gen.* **29**, 14, pp. 4019–4034.

Moody, R. V. and Patera, J. (1991). Discrete and continuous graded contractions of representations of Lie algebras, *J. Phys. A: Math. Gen.* **24**, pp. 2227–2258.

Mukunda, N. (1967). Unitary representations of the group $O(2,1)$ in an $O(1,1)$ basis, *J. Math. Phys.* **8**, 11, pp. 2210–2220.

Nakamura, K. *et al.* (Particle Data Group) (2010). Review of particle physics, *J. Phys. G: Nucl. Part. Phys.* **37**, 075021.

Nikolov, A. V. (1968). Discrete series of unitary representations of Lie algebra of group $O(p,q)$, *Funct. Anal. Appl.* **2**, 1, pp. 99–100 (in Russian).

Pajas, P. and Raczka, R. (1968). Degenerate Representations of the Symplectic Groups. I. The compact group $Sp(n)$, *J. Math. Phys.* **9**, pp. 1188.

Pajas, P. (1969). Degenerate Representations of the Symplectic Groups. II. The Noncompact group $Sp(p,q)$, *J. Math. Phys.* **10**, pp. 1777.

Perelomov, A. M. and Popov, V. S. (1966). Casimir operators for groups $U(n)$ and $SU(n)$, *Nucl. Phys.* **3**, 5, pp. 924–930 (in Russian).

Perelomov, A. M. and Popov, V. S. (1966). Casimir operators for orthogonal and symplectic groups, *Nucl. Phys.* **3**, 6, pp. 1127–1134 (in Russian).

Perroud, M. (1983). The fundamental invariants of inhomogeneous classical groups, *J. Math. Phys.* **24**, 6, pp. 1381–1391.

Peskin, M. E. and Schroeder, D. V. (1995). *An Introduction to Quantum Field Theory* (Addison-Wesley).

Pimenov, R. I. (1959). Axiomatic investigation of space-time structures, in *Proc. III All-USSR Math. Congress, 1956* (Moscow, USSR), **4**, pp. 78–79 (in Russian).

Pimenov, R. I. (1965). The unified axiomatics of spaces with maximal motion group, *Litovski Mat. Sb.* **5**, pp. 457–486 (in Russian).

Rembielinski, J. and Tybor, W. (1984). Possible Superkinematics, *Acta Phys. Polonica* **B15**, 7, pp. 611–615.

Reshetikhin, N. Yu., Takhtajan, L. A. and Faddeev, L. D. (1989). Quantization of Lie groups and Lie algebras, *Algebra i Analiz* **1**, 1, pp. 178–206 (in Russian).

Rosenfeld, B. A. (1955). *Non-Euclidean Geometries* (GITTL, Moscow) (in Russian).

Rosenfeld, B. A. and Karpova, L. M. (1966). Flag groups and contractions of Lie groups, in *Proc. Sem. Vect. Tens. Anal. MSU* **13**, pp. 168–202 (in Russian).

Rubakov, V. A. (2002). *Classical Theory of Gauge Fields* (University Press, Princeton, USA).

Schwinger, J. (1965). On angular momentum, in *Quantum Theory of Angular Momentum*, edited by L. Biedenharn and H. Van Dam (Academic, New York).

Tseytlin, A. A. (1995). *On Gauge Theories for Non-Semisimple Groups*, hep-th/9505129.

Vilenkin, N. Ya. (1968). *Special Functions and the Theory of Group Representations, Translations of Mathematical Monographs, Amer. Math. Soc.*, Vol. 22 (Providence, Rhode Island).

Voisin, J. (1965). On some unitary representations of the Galilei group, *J. Math. Phys.* **6**, 10, pp. 1519–1529.

Yaglom, I. M. (1963). *Complex Numbers* (Fizmatgiz, Moscow) (in Russian).

Yaglom, I. M. (1979). *A Simple Non-Euclidean Geometry and Its Physical Basis: An Elementary Account of Galilean Geometry and the Galilean Principle of Relativity* (Springer-Verlag, New York).

Yaglom, I. M., Rosenfeld, B. A. and Yasinskaya, E. U. (1964). Projective metrics, *Advances in Mathematics* **19**, 5, pp. 51–113 (in Russian).

Zaitsev, G. A. (1974). *Algebraic Problems of Mathematical and Theoretical Physics* (Nauka, Moscow) (in Russian).

Zeiliger, D. N. (1934). *Complex Line Geometry* (GTTI, Moscow, Leningrad) (in Russian).

Zheltuhin, A. A. (1985). *Para-Grassmann Generalisation of Superconformal Symmetry of Charged Fermionic String Model* (Preprint CPTI, 85-38, Moscow) (in Russian).

Index

CPSIA information can be obtained
at www.ICGtesting.com
Printed in the USA
BVHW042308211219
567403BV00005B/7/P